AF884888

Smart Money's Crypto Handbook

The Essentials of Cryptocurrency Investing

Phoenix Desmond

DeFi Magick
www.defimagick.com

Copyright © 2021 Defi Magick, LLC.

All rights reserved. No part of this publication may be distributed or reproduced in any form or by any means, without written permission from the publisher.

Disclaimer: This book is offered solely for educational and entertainment purposes. The author, publisher, and their affiliates are not offering it as legal, tax, accounting, or investment advice. All financial opinions expressed by the author aren't meant to persuade anyone to make a certain investment decision or follow a particular trading strategy. Profits and losses discussed in this book are based on hypothetical scenarios conceived using hindsight, and aren't indicative of actual results.

The author and publisher make no representations or warranties of any kind, and assume no liabilities of any kind with respect to the accuracy or completeness of the contents. They disclaim any implied warranties of fitness of use for a particular purpose. Neither the author nor publisher shall be held liable or responsible to any person or entity with respect to any loss or incidental or consequential damages caused, or alleged to have been caused, directly or indirectly, by the information contained in this book. Please use this information at your own risk.

TL;DR
Apply your own discrimination to the information in this book. Consider the opinions of the author as one of an infinite number of perspectives, NOT as absolute truth. Formulate your own opinions in order to create a strategy that works for you.

You are responsible for your investment decisions. Always be aware of the real risks of investing in any cryptocurrency. As a supplement to this book, it is recommended that you seek the advice of a financial or investment advisor before investing your hard earned money.

If you are interested in joining a thriving community of conscious crypto investors, and want to learn about the hottest coins that can lead to explosive returns, join the conscious crypto collective at defimagick.com

This book is dedicated to all conscious crypto investors–past, present, and future. May your choices help to usher in the age of true decentralization.

Contents

"It is well that the people of the nation do not understand our banking and monetary system, for if they did, I believe there would be a revolution before tomorrow morning."

–Henry Ford

You Are Playing an Elaborate Game

A crypto investor who was excited about his gains approached a saint and said, "I am up 50% on my investments this year!" The saint shook her finger at him and said, "Unless you can feel the same way when you are down 50%, you shouldn't be playing the game!"

In the **cryptosphere**, long term success is much more than a stroke of luck. It requires a unique skill set that can be mastered by those who are willing to cultivate patience, discipline, and a beginner's outlook. Crypto investors with the highest chances of realizing exponential gains are not only well informed, but also able to keep a steady frame of mind under all circumstances. Whether they are up 8x or down 70% on their investments, life won't skip a beat.

Whereas the internet has become the primary technology for sharing information, cryptocurrency will be the supreme technology for transferring capital (and much more). At this critical time in the history of money, those who act from a deep inner knowing of this truth are on the leading edge of finance. They are **conscious crypto investors**:

brave early adopters who aren't afraid to expand their horizons to both discover and invest in groundbreaking cryptocurrencies.

Conscious crypto investors cast their vote for projects with solutions to real-world problems that will disrupt legacy financial structures. Their absurd risk tolerance in the eyes of **nocoiners**–those who don't hold any cryptocurrency–stems from their faith that it will continue to transform the world in remarkable ways for generations to come.

This book is for people who want to steadily grow their savings over many years and decades, not watch it depreciate while banks loan it out for a hefty profit. Independent, self-motivated people who want to wean themselves from debt, and can clearly see the benefits of becoming their own sovereign bank. Reasonable, open minded people who see crypto as much more than a speculative bubble, and are looking for the safest, most effective way to take part in a financial revolution. Does this sound familiar?

This book is not for get-rich-quick fanatics who are willing to bet their life savings in one fell swoop. It's not for those whose rigid thinking creates "end all be all" scenarios. It's not for the faint of heart with a casual interest in cryptocurrency, or cynics who still question its validity. Finally, it is not for those who lack the desire to truly understand blockchain technology, or its massive global implications.

Swimming in the ocean is light and fun for someone who knows how to swim. But if you don't know how to swim, you can easily drown under the waves. This book is split into five main parts, which when taken together will help you to smoothly navigate the unpredictable waters of crypto.

Part 1 approaches money as a language of value that has become increasingly sophisticated over several millennia. We will break down some hard truths about banks, and offer an alternative view of cryptocurrency as an internet for sharing value using a platform of trust. Both the advent of Bitcoin as the first cryptocurrency, and its underlying technology will be explored in detail. The most common myths that seek to defame crypto will be debunked. Finally, you will learn important truths about how the crypto games work, and how smart money thinks.

Part 2 will build the foundation needed for making sound investments in cryptocurrency. It explores doing your own research, fundamental and sentiment analysis, and the different types of crypto categories to consider. You will also learn a method to calculate how much to reasonably invest, the different types of exchanges and how to use them, and the different types of orders and how to place them. Finally, we will consider various trading and investment strategies, including important rules for conscious crypto investors.

Part 3 simplifies the process of technical analysis. You will learn how to view charts, how to read candles, how to connect price and volume, apply moving averages, and spot trends to be able to make trading decisions. There is also a deep dive into corrections, the four stages of every market cycle, and how to trade in rhythm with cycles.

Part 4 is dedicated to securing your crypto investments. This includes how to browse the internet privately, and how to add layers of security with a standalone password manager and 2FA. We will explore private and public keys, and compare the different types of crypto wallets, including how to set up a software wallet. Finally, you will learn how to earn passive income on your crypto investments using staking and lending.

Part 5 covers the Tao of crypto trading, with philosophies to stay sane in the Wild West of crypto. Modern trading strategies will be explained in terms of ancient spiritual principles. This section also includes practices to help you harness your emotions so that they work for you, keeping you level-headed and disciplined in even the most adverse circumstances.

Aside from making money as an investor, what subconsciously draws many to cryptocurrency is the potential of a truly decentralized, free market economy that creates trust and cooperation. This book speaks to a growing number of conscious crypto investors who are not only committed to steady financial growth to support the immediate family, but also approach investing as a spiritual endeavor that will ripple out to the larger global community.

This book will give you the essentials to help you make sound investments in cryptocurrency. The information encompasses a balanced, methodical approach to investing that is transformative by its very nature. May the chapters that follow help you to both discover and apply the secrets of smart money in order to create the greatest positive change in the world.

Part 1
A Disrupter Is Born

Chapter 1

Money: A Language for Value

Money is one of the most important aspects of our lives, and its pursuit takes up a significant amount of our time and energy. It is also a critical social construct, since we create bonds whenever we use it. You could say that money is a literal extension of our vital energy. Ironically, it also happens to be a somewhat taboo subject outside of the home that is only vaguely understood. Aside from learning how to count it in school, we aren't taught about its deeper meaning, or how to manage it properly.

Money has been an integral aspect of every civilization. It is an ancient technology: a special language that allows us to express value. In fact, the language of money is even older than writing. During the Stone Age, money took the form of shells, beads, and feathers. The oldest hieroglyphics, which were the first forms of writing, reveal ledgers that kept track of money trading hands.

In ancient times, money was anything that had immediate value. People could barter items that could be consumed for sustenance: livestock, produce, grains, etc. From there, it took hundreds of thousands of years for money to take its first step toward being an abstraction. Precious stones and metals weren't given a perceived value by the collective until the Egyptians came along. Even though you couldn't eat or drink them, they were universally admired for their beauty. Their inherent value came from their properties: they were hard to come by, couldn't be replicated, could be divided and transported, and were very durable.

A few thousand more years passed by, and another technological innovation emerged that took the abstraction one step further. People got tired of lugging their gold around everywhere they went and burying it in the backyard. They realized that they could deposit it with someone they trusted in return for a special piece of paper. That paper certificate could then be used as a substitute for gold.

Money took another giant leap forward, but it wasn't the smoothest transition. On one side of the fence were the purists, who were convinced the paper didn't have any value despite evidence to the contrary. Then on the other side of the fence were the futurists, who recognized both the necessity and practicality of using paper money that was backed by gold. It took a few hundred years for widespread adoption, but eventually.... well, you know the rest.

Fast forward to 1950, when Diners Club issued the very first universal credit card made of paper. You can imagine how confused people were trying to understand how a series of numbers on a paper card could magically pay for everything. While paying for goods and services on credit had been around in some areas of commerce since the 1920's, it still took decades for mainstream awareness and adoption. We had to adjust to the idea of paying off statements periodically, and learn all about credit card fees and interest rates. But we did, because it allowed for a greater degree of freedom and flexibility with money. Now we could entrust another party to tally the totals and send us a monthly bill. Hundreds of transactions could be reduced to a single statement.

You can see the many ways that abstractions of money have helped to simplify the language of expressing value. The technologies themselves are brilliant–the problem lies in their inevitable corruption over time as "trusted parties" let their greed get the better of them.

Chapter 2

Magic Tricks with Money

The Roman Empire, one of the largest and most advanced civilizations of the ancient world, is a prime example of corruption. Rome's power came from military conquest. The more wars they won, the more resources they were able to plunder. But Rome's greed and carelessness led to a spending problem, one that was outpacing their ability to gather resources. When Nero came to power in 54 AD, he immediately began to debase the Roman currency by reducing the amount of silver added to each coin. He managed to reduce it to 90% purity, which enabled him to mint more coins so that Rome could continue to spend frivolously.

Future emperors diluted the currency even more, reducing the silver content to only 50% by 217 AD. Rome began to experience hyperinflation, with the prices of commodities like bread doubling in less than a century. This eventually led to invasions, civil unrest, and trade complications that forced farmers to become self-sustaining.

Even though government revenue was at an all time low, government spending was higher than ever. Cities desperately tried to build walls and strengthen armies to defend against invaders, which led to a further devaluation in the currency to 2% silver per coin by 300 AD. Rome's greed and selfishness led to its ultimate demise.

Full Faith and Credit?

For many centuries, the US dollar was backed by gold, which meant that you could literally take your dollars and exchange them for gold. But in 1933, FDR's executive order 6102 declared it illegal for US citizens to "hoard" large amounts of gold. FDR believed that this would be the quickest way to increase the money supply, since hoarding gold slowed down the economic recovery. At the time, cash had to be backed 40% by gold. This meant that for every $1 in circulation, there only needed to be .40 of gold locked up in the vaults. Remember Nero?

At Bretton Woods in 1944, the US dollar was declared the world's reserve currency since at the time, the US had the greatest amount of gold in reserves. But only twenty seven years later in 1971, the dollar was officially divorced from the gold standard by Richard Nixon. Now it was only backed by the "full faith and credit" of the US government. Here is the treasury's official explanation: "Federal Reserve notes are not redeemable in gold, silver or any other commodity, and receive no backing by anything. Redeemable notes into gold ended in 1933 and silver in 1968. The notes have no value for themselves, but for what they will buy."[1]

Did you know that the **Federal Reserve** (Fed) is an independent central bank, one that can make decisions without approval from the federal government? Even though it is subject to congressional oversight, it is self-funding, its board members can serve terms that span multiple presidencies, and they can conduct monetary policy and regulate financial institutions such as banks. In other words, a *lot* of power is concentrated in the hands of a small number of people with private interests.

Money on Steroids

It is obvious that the more money that is added to the total supply, the less it is worth. This has the consequence of also raising the price of goods and services. When money is injected into the economy, it's like a short term adrenaline rush, since it encourages more people to consume. The problem is that printing money doesn't add an equal amount of value to back the increase in supply. This is in complete contrast to the principles of a free market economy based on healthy supply and demand.

It is estimated that 22% of the circulating supply (at the time of writing) was printed in 2020.[2] What's worse, the US printed more money in June 2020 than in the first two centuries after its founding.[3] The total inflation since 1913 is more than **2,500%**. This is significantly higher than Rome's inflation growth during its downfall. The Fed is printing money at an unprecedented rate, and it is showing no signs of slowing down.

Ultrasound Money

When it comes to money, **time preference** is the value we place on receiving cash now vs. some time in the future. To put it into context, think about what happens each time another round of stimulus checks are printed. Knowing that the value of the dollar is rapidly declining raises our time preference, creating an urgency to spend rather than save. While a higher time preference may appear to work in our favor in the short term (we accumulate more junk we don't need), long term we are digging our own grave.

Ultrasound money, on the other hand, has deflationary features that inspire us to delay gratification and plan for our future by saving. It lowers our time preference by incentivizing us to spend less and invest more. The prospect of long term gain naturally keeps excess consumption in check. Essentially, a lower time preference creates a more sustainable future for both people and planet.

Chapter 3

Banks: A Negative Sum Game

Unless you are either a dinosaur, or you avoid debt like the plague, you probably have a credit card. And that credit card likely has an annual interest rate between 15-25%. That works out to 1-2% interest each month, and that's not including processing fees of up to 3.5% per transaction. Many of us simply shrug off what we consider the inevitable inconvenience of using credit cards. But have you ever wondered where banks are getting this endless supply of money they so eagerly loan out to you for such a sizable profit?

The Red Pill Is Hard to Swallow

Fractional reserve banking is a policy issued by the Federal Reserve which allows banks to keep only a small reserve of customer deposits on hand at any given time. They are free to loan out the rest of the funds to generate income. Proponents argue that this helps to stimulate the economy by putting money that would otherwise "just sit there in the bank" to good use. The reserve requirement was 10% up until last year. But as of March 26, 2020, the fractional reserve requirement for banks was reduced to 0%.[1] In other words, it no longer exists!

Make no mistake about it: the Fed, and all banks operating under its policies, are legally allowed to lend 100% of what you deposit, regardless of the size or type of account you have. They will use your deposits to lend money back to you and everyone else via credit cards, loans, and mortgages, for up to 25% interest. With an average APY for a savings account at 0.04%, this means you are providing them risk-free loans so that they can churn an annual net profit of up to 24.96%.

That's not even including the credit card fees, fees to withdraw money, fees to keep a minimum account balance, fees to wire money, etc. These realities are common knowledge, readily available to the general public. The more money you deposit into your bank, and the longer you keep it there, the more they profit at your expense.

Hopefully these statements are ringing some alarm bells for you right now. Weren't banks meant to serve primarily as a safe storage facility for our money? What happens during a crisis, when everyone lines up at the bank to take out their money at the same time? They go bankrupt. Bank runs weren't just a thing during the Great Depression–they have happened many times in the past decade all over the world.[2]

Banks and Freedom Don't Mingle

Keeping money in the bank is a negative sum game. Fiat currency is losing value by the day, and what banks give back hasn't come close to covering those losses for over a century. Inflation, the shadow tax behind every purchase we make, has caused the dollar to lose **96%** of its purchasing power since the Fed was formed in 1913.[3] The game is rigged, and it's time to cut our losses and play a new one.

It is irrelevant whether the Fed and the banks it represents are malicious actors in a plot against humanity, or well-meaning chaps who have simply made a few critical errors in judgement. The numbers don't lie, and it is costing us the most precious commodity we have: time. Every time the Fed prints trillions of dollars, we lose time. Without our input or agreement, our salary loses value. The mortgage and car payment, our children's college loans, etc. will take longer to pay off. With each passing day, we have to work longer at the same job to earn the same amount of money we earned yesterday.

This is an assault on our basic human rights to life, liberty, and the pursuit of happiness. Some believe we must rightfully pay a premium to safely store our money. But if 100% of our deposits can be loaned out to others, who really owns the money? Do you feel safe not knowing what banks are doing with the fruits of your labor?

The Middle Way

The initial anger many experience after discovering the pitfalls of legacy finance is often followed by anxiety and despair. "What choice do we have?" many wonder. "How would we function without banks? How can we ever hope to own a house, or use a credit card, send money to friends and family, or pay our bills?"

What if there was a middle way, one that allowed you to keep your bank account? If you could actually put your money to work for you, earning exponentially more on your savings and using some of that surplus to pay off your existing debt? What if you could gradually eliminate your financial burden and have more time to do what you love? Would you do it? Of course you would!

Enter **cryptocurrency**, a form of digital currency that is secured by cryptographic code. A $1,200 stimulus check from April 2020 that was invested in ethereum on April 12, 2020 would have gained 1,300% in one year. Meanwhile, had you simply deposited the check into your bank account, it would have lost 2% or more of its value from inflation. Let's do some math: $16,800 in crypto vs. $1,176 in the bank. Which do you think is the more sensible way to store your money?

Chapter 4

A Different Type of Internet

Many people don't realize that the internet, one of the greatest inventions of the 20th century, originally emerged in the 1960's. It took *three whole decades* for its user base to grow to 70 million.[1] Even then, it was relatively unknown, and received its fair share of negative press.

Figure 1: This article, which was published by the Daily Mail on December 5, 2000, cites research that people were turning away from the internet and email because it was too cumbersome and impractical to use.

Back in the mid-90's, how many people actually understood the internet's myriad potential use cases outside of email? Most scoffed at the idea that it would ever replace snail mail, or reach widespread adoption. In 1997, journalists were predicting it would fail miserably because no one could find anything. While they were focused on its limitations, Larry Page and Sergey Brin were building Google.

Then there were people like Jeff Bezos, founder of Amazon, who recognized the innovation as it was happening, and used it to reach hundreds of millions more people. Who would have thought that an online bookstore would evolve into the e-commerce giant it is today? Bezos admits he often had to explain the concept of the internet to nearly all of his early investors!

Mass adoption was *a slow process*. The first web page appeared 20 years after the first email was sent over the internet in 1971. Despite people's willful ignorance, the internet grew to over 1 billion users in 2005, and 4 billion in 2019. Today a large majority of the population relies on the internet for nearly every aspect of life. Can you imagine where we'd be without it?

Cryptocurrency: Powered by the Internet

Digital money isn't anything new. PayPal first began accepting peer-to-peer electronic payments online in the 1990's. The first online banking service was offered by Wells Fargo in 1995, and banks have been using digital ledgers to track money for several decades now. Cryptocurrency, which first appeared in the form of Bitcoin in January of 2009, is programmable money built on an internet of value. Cryptocurrency requires the internet for its storage, security, and transmission.

If you draw comparisons between internet use and crypto adoption, as crypto analyst Willy Woo has done, you will find that in 12 short years, crypto has already surpassed the number of internet users over 30 years after it was invented.[2] In the 1990's, the internet grew at about 63% on average each year. In contrast, cryptocurrency is currently averaging about 113% annual growth!

These numbers are significant, because if crypto continues to grow at its current rate, it will have over 1 billion users by 2025. When this happens, what took the internet 4 decades to achieve will have taken less than half the time for the cryptosphere. Adoption drives up demand, which will without a doubt lead to an exponential increase in the price of bitcoin and ultimately thousands of other cryptocurrencies.

The Boat Is a Cruise Liner-GET ON!

Many people who watched the price of bitcoin grow more than 16x in 2020 believe they "missed the boat" with cryptocurrency. But what they haven't yet realized is that cryptocurrency is a global asset class that will continue to grow exponentially. Let's explore why.

As of May 2021 there were 203 million crypto users, or about 2% of the world's population.[3] *The number of users doubled in four short months.* With this in mind, would it be unrealistic to expect that number to grow to 4-6% of the population in another year? That would inevitably lead to a steady boost in demand for cryptocurrency.

Now consider that a large majority of banks and institutions are just now warming up to the idea of cryptocurrency. Will they really have a choice, when they manage funds that are provided by the people they represent? Enormous financial institutions such as J.P. Morgan, Goldman Sachs, and Grayscale have all made their active participation in cryptocurrency crystal clear. What's more, every day an increasing number of financial service providers such as PayPal, Visa, Square, and Venmo are making it easier to buy and hold crypto.

Next, let's compare different assets' **market capitalizations**: the price of the asset multiplied by the circulating supply.

Cryptocurrency	$2.14 trillion
Apple	$2.17 trillion
Facebook	$894 billion
Gold	$11.7 trillion
Real Estate	$16 trillion
United States Dollar	$30.4 trillion
Stock Market	$49.1 trillion
Bonds	$120 trillion

Notice that cryptocurrency currently has about the same market cap as Apple. Some may think that's pretty high already, but Apple is a technology company, while cryptocurrency is a global asset class. We could compare it to gold, which is about six times the current market cap of crypto, or the dollar, which is fifteen times larger, but these are still only individual currencies. There are thousands of cryptocurrencies and climbing. Saying that crypto will rival gold and the dollar would be a severe understatement.

Bitcoin's Humble Beginnings

Bitcoin was the original cryptocurrency developed in 2008, coincidentally during the Great Recession caused by banks' over lending and careless investing. While millions of people were losing their jobs and homes, a mysterious figure(s) known under the alias of **Satoshi Nakamoto** was hard at work implementing a solution.

In the whitepaper for Bitcoin, Satoshi recognized the need for a truly **peer-to-peer** (P2P), decentralized electronic payment system "based on cryptographic proof." In other words, a mathematically secure network that allows any two parties to transact with each other *without* the need for a third party. Bitcoin uses **blockchain** technology. A blockchain is a chain of blocks containing data that is hashed to prevent them from being altered. In the case of Bitcoin, the data takes the form of transactions between users.

The blockchain creates what is known as a **public distributed ledger** (PDL), which represents a complete record of all the transactions that have ever occurred on a network. Anyone can view the history of transactions using blockchain explorers, going all the way back to the **genesis block**: the very first block.

A **hash** is a unique fingerprint that identifies a block and its data. To keep the blockchain secure, Bitcoin hashes data from each block into a 256-bit number using an algorithm known as SHA-256. Hashes from different blocks are combined to form the blockchain. If the data inside a block changes, the hash would have to change as well, which would invalidate all following blocks.

Since the blockchain is publicly distributed, anyone can join the network and receive a full copy of the blockchain. **Nodes** are computers that join a cryptocurrency network. Nodes help to create consensus by verifying the accuracy of the blockchain. A majority of nodes must agree which blocks are valid. If a previous block has been tampered with, its hash would create a break in the chain. As the number of operating nodes increases, so too does the security and stability of the network. This creates a **trustless** environment, where nodes maintain consensus and prevent bad actors from tampering with the blockchain.

Mining is the process where computer nodes validate transactions on the blockchain to earn rewards in the form of freshly minted bitcoin. Miners must solve complex math puzzles to win the right to add the next block of transactions to the blockchain. Bitcoin has a hard cap of 21 million, making it a deflationary asset.

Satoshi Nakamoto created the concept of mining to serve two main purposes: 1. gradually add more bitcoin to circulation, and 2. incentivize continued growth in the network with block rewards. Miners continue to join the network from across the globe every day.

Proof of Work (PoW) is a decentralized consensus mechanism that requires miners to expend energy by running computer software to validate transactions. The effort in the form of time, resources, and electricity to solve the equations is the PoW that earns mining rewards. Mining is analogous to precious metal extraction, since it takes measurable effort.

Miners compete to increase their hash rate, which enables them to earn larger block rewards. It takes about ten minutes to mine a bitcoin block.

Bitcoin has a built-in difficulty protocol that automatically adjusts the difficulty of the puzzle every two weeks depending on how quickly blocks are added to the network. Satoshi Nakamoto mined the genesis block with the following headline embedded in the timestamp: "The Times 03/Jan/2009 Chancellor on brink of second bailout for banks." A clever use of a timestamp, wouldn't you agree?

The First Bitcoin Poker Event

At that time Bitcoin was obscure, mostly known to cryptographers, tech fanatics, and online gamers. By 2010, the price of 333 bitcoin was about $1. During the first online bitcoin poker event, the buy in was 55 bitcoin, or $3.3 million at the time of writing. There were only four participants, and the winner walked away with 1,000 bitcoin! [4]

One of Many Digital Gems

Bitcoin is both unique, and valuable for several reasons:

1. It is the first decentralized digital currency.
2. It is open source, with a public design that no one can own or control, but anyone can work to improve.
3. It is scarce, with only 21 million total bitcoin that can be mined. Close to 19 million are already in circulation. The remaining 2 million will be mined by 2140. As a deflationary asset, it's value will naturally increase over time.

4. Since it is fully decentralized and backed by mathematics, it is durable, immutable, borderless, and trustless. As long as the internet exists, there will be independent nodes that will continue to collectively maintain the integrity of the blockchain.

5. Unlike gold, bitcoin is divisible to the eighth decimal place. **Satoshis** (sats) are the smallest unit of bitcoin recorded on the blockchain. There are 100 million sats in 1 bitcoin. Sats are to bitcoin what cents are to the dollar, only much smaller. You can purchase a dollar of bitcoin with a few taps on your phone.

6. Bitcoin is also extremely portable, with the capability to be sent anywhere in the world in about ten minutes. Fees are minimal compared to those for wiring fiat currencies overseas.

Bitcoin has created much more than a simple store of value. It is a decentralized platform of trust: a template which can be applied to every other aspect of our lives. Currency is to blockchain what email is to the internet. Email is a critical internet application that inspired the advancement of internet technology. In the same way, the unification of existing systems of money is the first of countless applications to be developed using blockchain technology.

We can now use cryptocurrency to securely send either pennies or billions of dollars across the globe with complete transparency. Money is taking on a new dimension as it flows across the blockchain with the same speed and efficiency that information flows over the internet. The evolution of currency beyond the need for trust, permission, or borders has been the catalyst for radical innovation in the cryptosphere. Universal ecosystems are developing that shrug off the closed systems of the past in favor of open source collaboration.

Children today will never need bank accounts. Instead, they will download personal banking apps that will allow them to instantaneously transact with anyone in the world in seconds. They will be able to convert their money to any currency with a few swipes, and will scan their crypto wallets to rent cars, check in to airlines, access data storage, and listen to streaming music. The possibilities are endless!

Chapter 5

Crossing the Murky Swamp of Misinformation

An Iron Stomach for FUD

Did you know that Bitcoin has been declared dead over 400 times from 2010 - 2021? Check out 99bitcoins.com/bitcoin-obituaries/ and see for yourself. Nearly every mainstream media outlet and financial institution has, at some point or another, shamelessly bashed bitcoin. They do this by spreading **FUD**: **f**ear, **u**ncertainty, and **d**oubt. If you pay close attention, you'll hear the same narratives on repeat. It is a safe haven for money launderers; it requires more electricity than entire countries; it has no value since it isn't backed by anything; it is inefficient and the fees are insane; it is so volatile that it can never be considered a valid currency or store of value.

While it would take forever to counter every argument against cryptocurrency, some myths are worth debunking. You will in all likelihood have countless opportunities to enlighten your friends and family over the next 5-10 years. When they look at you cross eyed for offering an alternative to an economy that is hobbling on its last leg, these insights will perhaps help to shine some light into the darkness.

Myth #1

"It's a ponzi scheme-fake internet money that isn't backed by anything!"

Cryptocurrency is backed by mathematics, and the ever-expanding network of thousands of nodes that run 24/7, 365 days a year in every corner of the planet to maintain the integrity of the blockchain. The US dollar, on the other hand, was removed from the gold standard fifty years ago, and is now backed solely by "the full faith and credit of the US government."[1] The same government that is currently **$24 trillion in debt**.

We know that fiat money's diminishing purchasing power motivates us to spend it faster. Central banks celebrate its short-term stability, while conveniently ignoring it's steady debasement over years and decades. At the same time, they condemn the short-term instability of cryptocurrency but ignore its remarkable growth over the past twelve years. $100 invested in the dollar in 2009 is worth $80 today. $100 invested in bitcoin in 2010 (1,000 bitcoin) is worth $63 million today (at 63k per bitcoin).

It is plain to see why bitcoin has been the best investment of the decade. Despite opposition from governments and banking authorities around the world, bitcoin has risen **630,000%** in value in eleven years. What does that say about it's value to the people? Bitcoin is clearly ultrasound money.

Bitcoin has 19 million coins in circulation, a hard supply cap, thousands of nodes and growing, and ~1.7% of its supply minted in the last year.* Now compare this with the US dollar. There are currently 22 trillion in circulation, with an unlimited supply cap, one node, and 22% of its supply minted in the last year. Ready for the cherry on top? The top 1% of holders own over 30%! Which sounds more like a ponzi scheme to you?

*With ~52,500 blocks mined annually, a current reward of 6.25 BTC/block, and 18,781,693 BTC in circulation at the time of writing. (52,500 x 6.25)/18,781,693 = 0.0174, or 1.7%

Myth #2

"It will never reach mainstream adoption."

The argument that crypto will forever remain a speculative investment, doomed to slither in the shadows of legacy finance is perhaps the most senseless of the lot. Global crypto adoption measured by P2P retail transactions and trade volume on exchanges soared **880%** from July 2020 to June 2021.[2] In reality, institutions were the last domino that fell when legions of eager investors began to demand a piece of the action.

Did you know that financial powerhouses that control trillions of investor dollars, such as Morgan Stanley, Goldman Sachs, J.P. Morgan, Citi, State Street, and Wells Fargo, are all offering crypto funds to their clients? PayPal, Venmo, Mastercard and Visa also offer a variety of crypto-related services to millions of customers around the world. Add to that the fact that three dozen publicly traded companies currently own billions of dollars in bitcoin alone. Hundreds more are expected to follow suit as they move their capital into crypto as a safe hedge while the purchasing power of the dollar continues its destined decline.

Ironically, nearly every single one of these financial giants once publicly dismissed cryptocurrency as a worthless investment! There is one reason they have all made a swift 180 in recent years: **game theory**. If one major player decides to invest in cryptocurrency, the others will eventually follow suit. A critical mass of crypto-friendly retail investors are banging on the doors of the financial institutions who represent them. Even if the institution has a strong dislike for crypto, when other institutions make it part of *their* investment strategy, they feel pressure to do the same in order to stay competitive.

A 2021 Global Blockchain Survey by Deloitte Blockchain found that 76% of finance professionals think crypto will become either an alternative to, or replacement of fiat currencies in 5-10 years.[3] Among 1,300 executives polled, 81% think that blockchain has already achieved mainstream adoption, and that it is **scalable**: it has the potential to become faster and more efficient. What's more, 73% think institutions should adopt crypto in order to keep up with a "seismic shift" towards cryptocurrency.

Thousands of banks are scurrying behind the scenes to keep their clients' money in house by creating crypto gateways, while social media platforms like Twitter are integrating their own crypto wallets. Over fifteen companies, including NYDIG, Fidelity, First Trust, VanEck, and WisdomTree, have already filed for a bitcoin **Exchange Traded Fund** (ETF). An ETF is a group of securities, such as bonds, stocks, commodities, or currencies that trade on a public exchange. For example, the SPY ETF consists of all the stocks in the S&P 500 index.

ETFs are popular among investors since they can encompass hundreds, or even thousands of securities. They create a bridge between legacy finance and the cryptosphere. Investors who don't want to deal with the technicalities of buying and storing cryptocurrency can gain exposure by purchasing shares of an ETF instead. As of 2021 there are over 8 trillion dollars invested in ETFs globally. If even a fraction of that flowed into cryptocurrency, it would be a shot in the arm for the entire cryptosphere.

Myth #3

"It's too volatile to ever be a currency, let alone a store of value."

Even after over a decade, cryptocurrency is still a toddler compared to every other market. Gold, for instance, has been around for millenia. During this time it has slowly, steadily earned its value through the collective consensus. And while Bitcoin has already proven its scarcity, durability, and universal value, it has only been adopted by a small fraction of the population.

Think of the cryptosphere as a young child. Children can be exuberant for hours, but then just when you think the party will never end, they suddenly burn out and throw a temper tantrum. Anyone who has ever parented a two year old is intimately aware of this cycle, which repeats itself at least once a day.

Once children grow up, they are less prone to extreme swings of emotion, since they are better able to manage their energy. Their attention span is more than twenty seconds, and they don't need to nap as much. But if as a parent, you become hyper focused on a young child's temperamental habits, and unable to recognize his or her slow, steady maturity over the years, then you won't have faith in the child's upward trajectory. In other words, your time preference is too high–you are expecting to receive the rewards of parenting too soon.

The moment you take a deep breath and zoom out, you see that despite your child's sporadic tantrums (volatility), over time her/his behavior is gradually improving. This expansive view automatically lowers your time preference, since it becomes much clearer that your relationship is headed for greener pastures in the years and decades to come.

In Figure 2 below, you can see that in just over two months, the price of bitcoin fell ~56%. If this were the only information used by investors, no one in their right mind would want to put a penny into crypto.

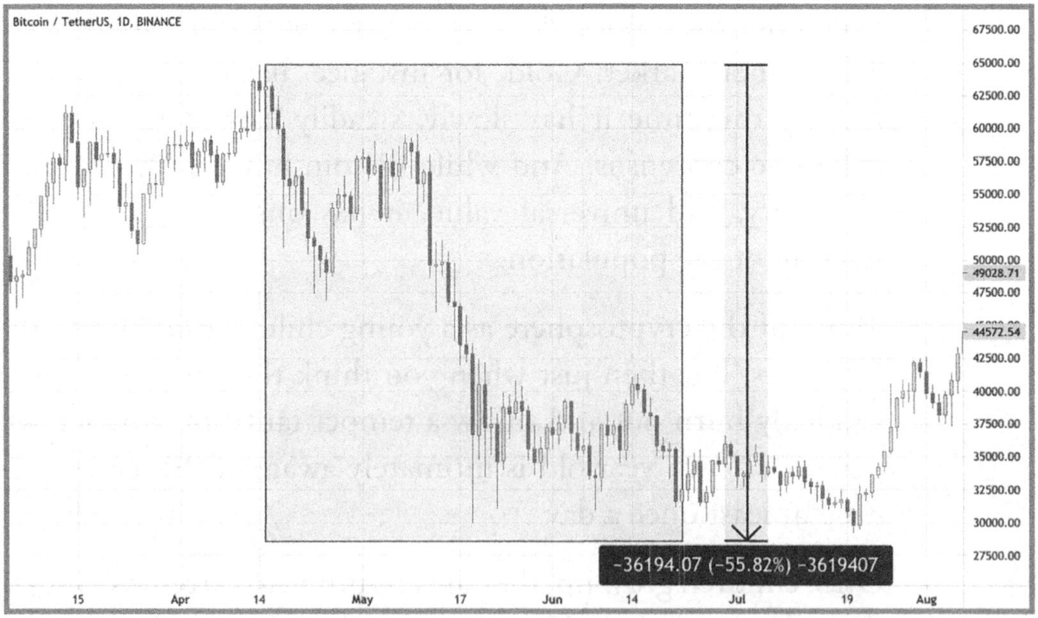

Figure 2: Bitcoin fell ~56% from April 14 - June 22, 2021.

Now let's zoom out, including price data from March 2017 to August 2021 in Figure 3. Looking at the chart, it is abundantly clear that bitcoin is a sound investment for those with a lower time preference. In other words, anyone who is willing to parent the unruly child named bitcoin is destined to receive exponential rewards in the years to come. Once you understand blockchain technology and its implications, short term volatility becomes an afterthought.

Figure 3: Bitcoin from March 2017 to August 2021. The blue box shows the same price data as in Figure 2 above. Perspective is everything!

Now let's take the volatility debate a step further. Imagine for a moment that you live in a country like Venezuela, which experienced 2,355% inflation in 2020. Even though this is an extreme case, in 2020 there were 20 countries with an annual inflation that was higher than 10%.4 If you lived in one of them, and you saw Figure 3, would you be the least bit concerned about bitcoin's short term price swings? It is no coincidence that several of the countries with the highest inflation are also at the top of the list for global adoption in 2020. Especially when you consider crypto's overall performance vs. every fiat currency over the last decade! Generally speaking, western nations lack the urgency to move capital into crypto because they haven't directly experienced the level of financial instability that other countries have. At least not yet...

It will likely take several more years for cryptocurrency to mature. During this time we can expect more frequent and significant price swings, since the market cap of most coins is still relatively small. But this shouldn't dissuade you from investing. With greater risk comes a higher potential for reward. One day Bitcoin will have such a high market cap that it will barely move a quarter of a percent, and we will think back to those 30% swings with a tinge of nostalgia!

Myth #4

"Energy consumption is outrageous. Miners will be shut down, and Bitcoin will die."

The Bitcoin Mining Council was founded shortly after Elon Musk expressed concerns about Bitcoin mining energy usage. In their first press release, they estimated that the total sustainable power mix for miners is as high as 56%.[5] This would make Bitcoin one of the most sustainable industries on the planet, far surpassing most nations. The US, for example, uses only 30.5% sustainable energy, and China clocks in at a measly 15%.

The chart in Figure 4 reveals a stark contrast to all the outrageous claims about Bitcoin mining energy usage. Mining only utilizes 0.117% of the world's total energy use!

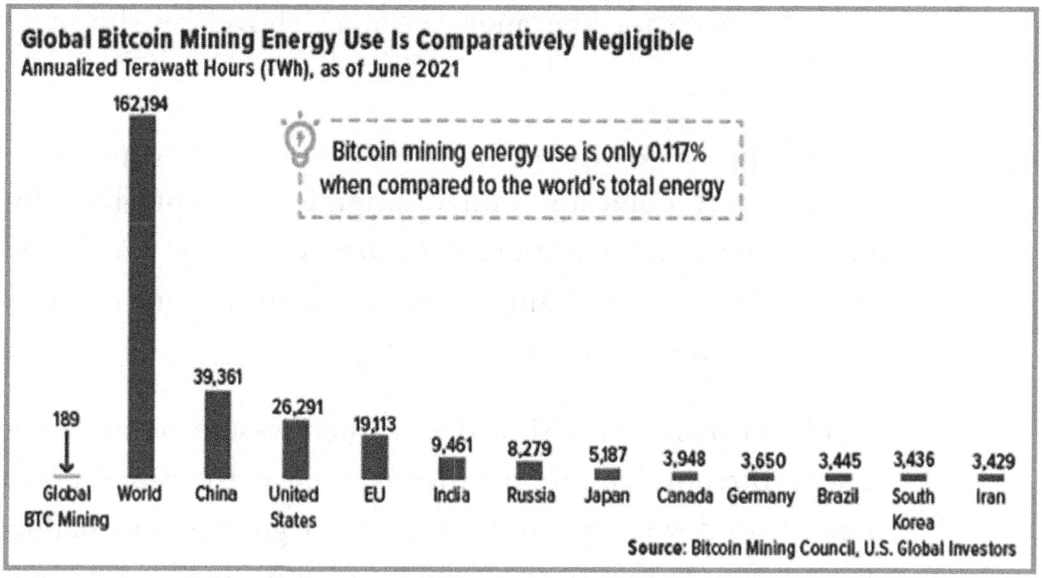

Figure 4: 2021 global Bitcoin mining energy use according to Bitcoin Mining Council.

China has been trying to ban Bitcoin since 2013, and for good reason. It is a threat to the digital yuan, China's newly introduced **central bank digital currency** (CBDC). The digital yuan and other CBDCs will allow governments to track every possible transaction.

There are many who mistakenly believe that CBDCs will somehow make cryptocurrencies irrelevant. While CBDCs may utilize blockchain technology, they definitely aren't decentralized. They basically take all the limitations of fiat and repackage them for the digital world. This allows for unprecedented levels of surveillance and control. The trustless, permissionless, borderless environment created by cryptocurrencies will be a foreign language to CBDCs.

China's authoritarian regime has been remarkably successful at censoring its citizens and stamping out drug trafficking and organized crime. Ironically, Bitcoin has been the velveteen rabbit that always manages to escape the angry farmer's shovel by slipping under the fence.

What happens when you ban transactions and criminalize mining in one country? Logically, miners migrate to crypto-friendly countries that welcome them as a boost to their economy. We know because it already happened in China, which effectively shut down 60% of the Bitcoin network in a few short weeks.

If 60% of banks were shut down, there would be mass hysteria. Not so with Bitcoin. Despite the inconvenience, the Bitcoin network experienced no downtime, and no need for government bailouts. Miners simply relocated their equipment and continued validating blocks.

Developing countries will always be incentivized to recruit miners as a means of generating income from otherwise wasted energy. El Salvador is already exploring the possibilities by harnessing volcanic energy to mine bitcoin.[6] Consider other rural areas where geothermal, hydroelectric, solar, and wind energy can be generated in abundance, but can't easily be transported long distances to cities. Also consider the fact that renewable energy isn't alway needed when it is generated (mid-day), causing it to go to waste. Supportive legislation will allow miners to soak up what would otherwise be inevitable losses from renewable energy systems, thereby encouraging their rapid expansion.

How can anyone hope to hunt down a decentralized technology with a global network that is growing faster than the internet, and no central figurehead to strangle? The cryptosphere is the digital embodiment of Hydra: cut off one head, and nine more will grow in its place.

Myth #5

"It's dirty money used by criminals."

Since cryptocurrency is fully decentralized, this also means it is the most traceable currency ever known! Why would criminals want to deal in a completely transparent currency, where all their transactions can be tracked by highly sophisticated blockchain analytics firms such as Chainalysis, CipherTrace, and Coin Metrics?

Cash is the original P2P, private form of money that carries the golden seal of approval for the most elite criminals on the planet. Governments are stepping up their efforts to gradually replace paper money with digital payment networks, since an overwhelming majority of illicit transactions are still carried out in cold hard cash.

The Bottom Line

As long as people value their time and freedom, bitcoin and other cryptocurrencies will do more than simply survive–they will thrive. Do not allow FUD to deceive you! FUD is market manipulation at its finest, and is designed to trick you out of your money. Ironically, most exchange outflows happen soon after a big crash. In other words, when it looks like everyone is selling, institutions are buying hundreds of thousands of bitcoin and moving them to external wallets for safekeeping.

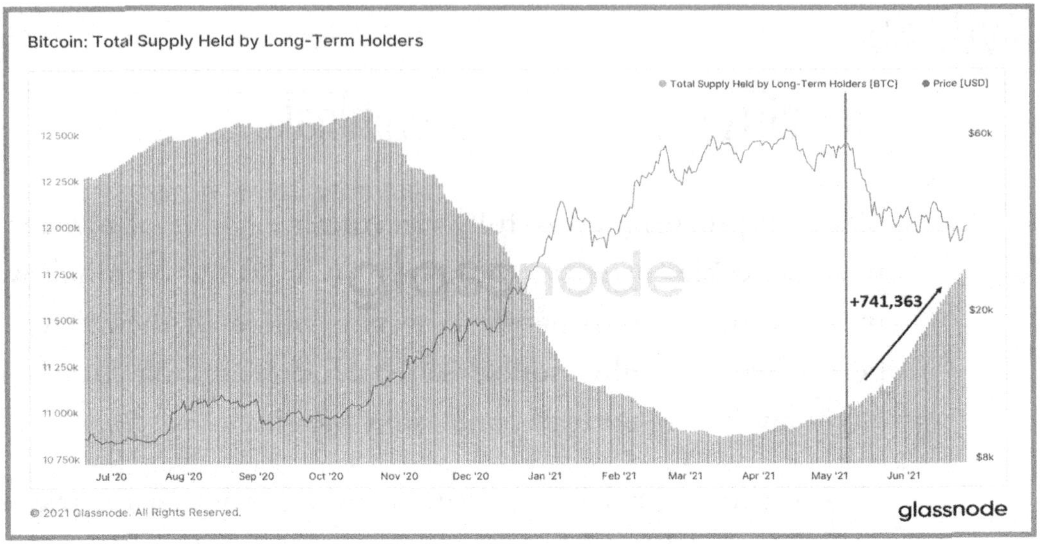

Figure 5: While the price of bitcoin tanked in May of 2021, the amount of supply by long term holders steadily increased by 741,363 bitcoin. Data from Glassnode.

Media is primarily driven by hype and controversy. This is especially the case when it comes to new concepts that are confusing to the masses. When the automobile first appeared, most people were convinced that it would never replace the horse and buggy. It was actually considered a moving fire hazard! Horse and buggy fans proclaimed cars were unreliable and inefficient. They said there was no sense buying cars since they eventually would run out of fuel, and there was no way they would create enough demand for fueling stations. Now, when was the last time you saw a horse trotting down the highway?

Clickbait efforts step up whenever crypto reclaims the spotlight during bull runs. Don't be discouraged the next time the FUD clowns come moping around. Cryptocurrency was forged in the fires of disruption–it is designed to endure. Media narratives are trivial in light of all the exciting new developments in blockchain technology every day. When that becomes your main focus, you won't be deceived into thinking crypto is a passing trend ever again.

Chapter 6

Cryptocurrency's Eccentric Following

Part of what makes cryptocurrency so interesting is the broad spectrum of people it attracts. Originally crypto was limited to mostly purists: super early adopters who have an excellent grasp of blockchain technology. They may have pitched Bitcoin to you back in 2010 with so much excitement that you thought they were part of a digital cult (back then, they were). Unfortunately for many of us, we dismissed them as manic speculators and continued about our lives as if nothing ever happened.

Purist investors are sometimes referred to as **maximalists** (maxis): those who are convinced of the superiority of original cryptos such as Bitcoin and Ethereum. To date, they have seen the largest returns, and they deserve them for sticking it out through three bear markets. You wouldn't be able to swipe their bitcoin from their cold, dead hands. Hopefully they trust their children enough to at least share their private keys! (more on that later)

Today the cryptosphere encompasses a much broader spectrum of users from all over the globe, including all ages, races, and beliefs. There are conspiracy theorists, hip TikTok celebs, large net worth individuals, adrenaline junkies who get a quick fix from a market that never sleeps, get rich and quit your dayjob hope fiends, and everyday people who simply want something that is more fair and efficient. Because of crypto's inclusivity and power as a great equalizer, a significant number of users are actually unbanked people living in third world countries.

Why the Unbanked Are Flocking to Crypto

Did you know that approximately 31% of the global population–2.4 billion people–are unbanked? [1] In every continent except North America, Europe, and Australia, a significant majority of people don't even have bank accounts. There are several reasons for this:

1. There is no existing financial infrastructure. Banks are capitalistic, not altruistic. They won't take the risk to expand into areas where they don't think they can turn a profit.
2. Where financial services do exist, they are too expensive or impractical for people to use, sometimes even charging negative interest rates. Imagine paying fees just to keep your money in the bank.
3. People don't have the required personal identification to open a bank account, often because of government inadequacies.
4. Many people in unbanked countries are already suffering from government corruption and hyperinflation, leading them to mistrust financial institutions.

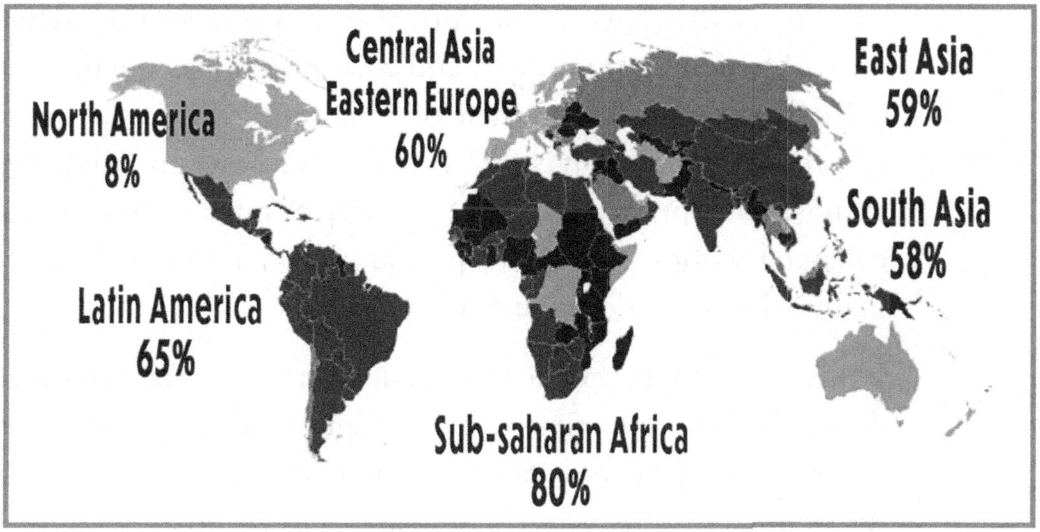

Figure 6: Percentage of adults who are unbanked by region. Notice the huge disparity between North America and every other geographical region.

Why does this matter? Imagine a hardworking farmer who needs a loan to lease a plot of land for his crops, but he is completely out of luck because he can't prove his creditworthiness. Or consider a migrant worker who is forced to pay an average of 7% in fees just to send the money she has earned to her family overseas, which then takes up to five business days to reach them. In 2020 alone, such remittances to low and middle income countries ($540 billion total) cost **$37.8 billion.**[2]

Even those who simply want to put money aside each month so that their children can one day receive a college education have no recourse aside from hiding it and praying. These are serious financial obstacles that hinder a majority of the world's population on a daily basis.

Now consider what happens when those same people begin to realize that they can send payments in seconds at a fraction of the cost (sometimes pennies), 24/7, 365 days a year, with 100% uptime and no one to freeze funds for any reason. Without a formal identity or bank account, with a few mobile apps they can protect themselves from the woes of hyperinflation, theft, and corruption. Despite the fact that their fiat currencies are losing their purchasing power by the hour, they can be proactive and safely store their capital. Not only that, they can also exercise the option to earn a fair interest rate, or borrow against their collateral without having to pass any special checks. Believe it or not, many are already doing just that.

If this all sounds far fetched to you, let's start by taking a moment to break down the ten countries with the highest adoption of cryptocurrency according to data from Statista. These are ranked by the percentage of the population that actively buys and sells crypto.[3]

1. Nigeria: 32%
2. Vietnam: 21%
3. Philippines: 20%
4. Turkey: 16%
5. Peru: 16%
6. Switzerland: 11%
7. India: 9%
8. China: 7%
9. USA: 6%
10. Germany: 5%

Notice that the wealthier countries have significantly lower adoption, with only four first world countries making the top 10. How can this be? Why would people in poorer countries be rushing to invest in a highly volatile asset class? Is it that our financial institutions are less corrupt, or have we simply become complacent with an antiquated banking system?

How do less developed nations without access to banks buy and sell crypto? Over ⅔ of the unbanked have access to the internet via a smartphone. Even people in the poorest rural regions without access to reliable electricity or running water have smartphones. In some areas, people will pool their resources to invest in a single solar panel capable of charging several phones simultaneously. Examples of this would be farmers that rely on weather feeds for their crops, families who use them for basic communication, or small business owners who need to access e-commerce.

DeFi: Goliath's Worst Nightmare

If you want to know all the wonderful ways the unbanked (and the banked) already use crypto to their advantage, you need to understand the basics of **Decentralized Finance**, or **DeFi**: a decentralized network of financial applications that run on the blockchain. DeFi creates the digital infrastructure that allows for the seamless storage and transfer of assets. These **decentralized applications**, referred to as **Dapps**, are like an interconnected web of money legos with the following traits in common:

1. Like the internet, they are open-source, so that anyone can apply the code to build their own Dapp. Open sharing of protocols greatly accelerates innovation.

2. They run on **smart contracts**: programs that are only executed after specific parameters are met. Smart contracts are mathematical functions that are censorship resistant, and don't require trust or permission.

3. They remove the need for third-party intermediaries, and the inherent corruption that they breed.

4. They are borderless and permissionless, which means they can be used by anyone, anywhere, anytime, independent of identity or creditworthiness.

Anyone who owns any amount of crypto can leverage the incredible power of DeFi. Even without a bank account, it is possible to use cash to buy bitcoin and other cryptocurrencies at thousands of **Bitcoin ATMs** (BTM) all over the world. There are several hundred in the US alone. There are also trading platforms such as Paxful and LocalBitcoins, which have several options for P2P trading of bitcoin and other cryptocurrencies that don't require a bank account or personal identification. What's more, you can now use any method of payment to buy crypto.

Once you have purchased cryptocurrency–whether or not you have a bank account–you could send it to your family for a fraction of the cost of a remittance. Fast, efficient, cross-border transactions that are safely verified by network participants in minutes, whether you are sending it across town or 10,000 miles overseas. Your family could then safely store the crypto, use it to buy goods and services, or cash it out into fiat using the same methods above. Are you beginning to see the implications here?

Payment processing and money transfers are only the tip of the iceberg. If you wanted to store your crypto instead of sending it to your family, with a few taps you could lock your coins on the network to earn rewards, loan them out and receive daily interest using a decentralized lending platform, swap them for other coins on a decentralized exchange, or use them as collateral to take out a loan. This is all done from a simple smartphone, using Dapps created by the people, for the people. No need for banks, proof of nationality or other legal documentation.

This all may seem like some kind of infernal sorcery right now, but it will become crystal clear in the chapters to come. DeFi is an integral aspect of cryptocurrency, one that is cranking up the control knob on the velocity of money to max speed. The ability to transfer value quickly and efficiently is a hallmark feature of any productive financial system. DeFi offers us a truly free market economy characterized by efficiency and equal opportunity.

Chapter 7

Welcome to the Crypto Games

The cryptosphere is an elaborate game. New players enter the market every second from every corner of the globe. At least initially, most have the sole aim of turning a quick profit. Although some will come and go relatively quickly, the majority will linger out of sheer curiosity. Money may be the hook point, but crypto's potential as a technology will keep them coming back for more.

Dot-com Nightmares on Repeat

It is common knowledge that 9 out of 10 businesses will fail. A crypto venture is a business like any other. Projects emerge with lots of hype and fanfare. While some will stick around, most will fade into the background.

Remember the dot-com bubble? People were eager to throw their money at any stock that boasted they were an internet-based company. When the curtain was drawn, and it came time for these companies to deliver on their promises, most of them tanked.

The crypto craze shares many similarities with the dot-com bubble. Investments are mostly speculative, since a significant number of crypto projects have yet to deliver a technology that can be fully implemented. What's more, among those that are in use, none are immune to bugs, hacks, and FUD attacks. What's trending today can easily be replaced tomorrow by a superior technology that solves more problems with greater efficiency.

Remember when AOL dominated email, internet connectivity, and internet news in the 90's? When dial up modems became obsolete, AOL fell out of favor and went from a 200 billion market cap to being sold to Verizon for $4.4 billion.[1] Since then Google and Yahoo have replaced it with far more advanced technology.

In the cryptosphere, projects with brilliant developers who are backed by well known investors and hyped by thriving social media communities can still fail. Meanwhile, coins such as DOGE, with no real use case and minimal development, can skyrocket when hyped by billionaire pranksters. Similar to the internet, winners and losers in the cryptosphere won't be immediately obvious.

Understand what you are getting yourself into before you decide to invest in the cryptosphere. Cryptocurrency is the evolution of the Wild West, only now bitcoin is being mined instead of gold. It's a daily stroll through the virtual streets of Vegas, where you must use your wits to uncover threads of integrity in a den of thieves. Does it mean everyone involved is greedy and corrupt? Definitely not! But as with any budding market, separating the wheat from the chaff is a process. It will take several years for the hysteria around crypto to quiet down and be replaced by lasting stability. In the meantime, this book will inspire you with the tools to discover the most meaningful projects with the highest potential.

Wolves Devour Sheep

It is said that about 10% of traders in the stock market will make money. How is this possible? Believe it or not, there is a one word answer: impatience. Warren Buffet is known to have said: "The stock market is a device for transferring money from the impatient to the patient." While Buffet isn't exactly a fan of crypto, his wisdom is extremely relevant to the cryptosphere when you consider how far we still have to go.

Let's get real for a moment. The average retail investor will throw money at anything that appears to be increasing in value. They likely won't pay much, if any attention to the age of the project, what the team behind it is creating, what it's chart looks like, etc. Money is the driving force behind all markets. Crypto is no exception.

In this game, there are **sheep** and there are **wolves**. Sheep are unwilling and unable to think for themselves, and follow both the opinions and decisions of the herd. They are easily manipulated by emotions such as fear and greed, since they haven't learned how to control them. They therefore fall prey to market fluctuations.

Wolves always enter the game knowing their strategy, and they don't allow emotion to lead them astray. When they act on their plans, they are swift and decisive. They are masters at blending in with their prey, keeping their presence unknown while maintaining proximity.

Patience is a wolf's mantra. They follow through on their commitment as if it were a matter of life or death. They may work together, but they are perfectly OK going solo if and when necessary. In other words, they know how to think for themselves.

Wolves eat sheep. The next time you look at a chart, pay close attention to the sharp peaks. Whenever you see a quick reversal in price to the downside, a pack of wolves gathered and pounced on the opportunity to sell their bags to unsuspecting sheep. Even though the turning point can take place in the span of several minutes or hours, the wolves have been ever present, preparing for this moment for weeks, months, or even years.

When you see the price climbing steadily on a chart, sheep were growing their flock while wolves watched from a distance. As it reached the top, the momentum influenced the herd to continue to climb despite the impending danger. Had they stopped to think instead of blindly following the herd, they might have averted the catastrophe.

As opposed to nature, in the cryptosphere wolves can't eat sheep who wait patiently for the bloodbath to end. In other words, you don't lose money until you sell. You can be down 400% on a coin in a bear market, but if you wait until the bulls return, that same coin could retrace its steps and rise 800%. This would transform your 400% loss into a 400% gain. Of course there are no guarantees, but *most money is lost from impulsive decisions that take place either shortly before or after a crash.*

Both the sheep and wolf have their respective place in any financial market. But these categories aren't always cut and dry. People may frequently move between the two ends of the spectrum depending on the day, their mood, and market conditions. Your top priority is to master your emotions by recognizing and uprooting your sheepish tendencies that arise in different situations.

People are fickle, and wary of long term commitments. Harsh but true! If they weren't, we would see slow, steady growth in the price of all cryptocurrencies, without any major hiccups along the way. Just looking at the 2017 bull run, bitcoin experienced 7 major corrections that averaged -33%.

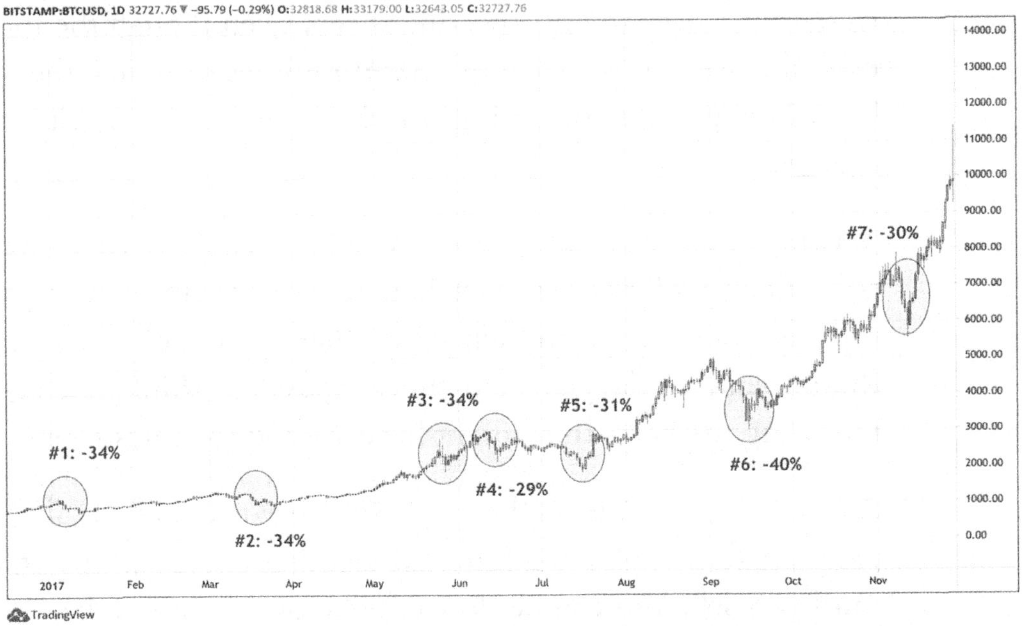

Figure 7: Seven corrections of 29% or more occurred during the 2017 bull run.

Bitcoin has the largest market capitalization. Theoretically, this means it should have the lowest volatility of all cryptocurrencies. But even bitcoin isn't immune to **pumps and dumps**: rises and falls in price over time.

The truths about wolves and sheep and the volatility they create aren't meant to scare you off from investing in the most promising technology of the century. Instead, they are disclaimers to keep you from getting wrapped up in hype cycles. Smart money is like a cautious driver that can anticipate dangerous situations, and therefore avoid them altogether. A sheep is like a reckless driver swerving in and out of his lane while texting his girlfriend. Even though they are destined to cause trouble for unsuspecting drivers, if you know what to look for, you can keep a nice safe distance.

Chapter 8

Smart Money = Your Highest Potential

Big money isn't always smart money, and smart money isn't always big money. What makes money smart isn't how much money is involved, but how it is used. One hundred dollars invested in bitcoin in 2010 was a smart money move. The same hundred dollars invested in bitcoin at the peak of the 2017 bullrun was a dumb money move.

Dumb money refers to an impulsive attitude that can be seen in those who are often scattered in their thinking and extreme in their emotions. Dumb money will act before thinking, making uninformed investment decisions that maximize risk and the potential for loss. This way of being is commonly found among gamblers and addictive personality types. Identifying and eliminating dumb money habits is the first step toward becoming smart money.

The Alchemy of Smart Money

Smart money is not a fixed title that is limited to an untouchable elite class. It is universally applicable, and transient since it always has the potential to become dumb money. We can all think of times we've made smart purchases, and *really really* dumb purchases. For example, imagine a car that you bought that spent more time in the shop than on the road. From the outside, the car may have looked flawless, but one look under the hood would have revealed otherwise.

Even though you may have lost a significant amount of money on a junk car, let's imagine one way you might have changed your fate in the long term. Had you been able to recognize the breakdown trend, you could have sold the car at a loss and invested the money in an incredibly reliable electric car that would save you thousands of dollars in gas and maintenance over the next 5-10 years (hint: Tesla). The money you save has the potential to even cover the sticker price of your car. In this case, a simple decision can instantly transform dumb money into smart money.

Even the richest people on the planet have made plenty of dumb money mistakes. Every decision that you make about money is an opportunity to redeem the mistakes of the past, and thereby create a brighter future. The trick is to learn from our actions without harboring regret. We also want to study and emulate the tendencies of smart money over time. These habits are integral to the success of any venture, and are especially critical in the cryptosphere, where minor errors in judgement can cost you a fortune overnight.

Happy First, Rich After

Understand that the rich minority–those who are able to consistently grow their wealth over years and decades–aren't always the happiest. Sure, money can help to create a more comfortable life that adds to happiness, but if there is a lack of balance between family life and career, work and play, then we pay the price of our mental well being.

Those with true wealth aren't simply rich–they are abundant in every aspect of their lives. They live a healthy lifestyle, enjoy a broad circle of friends, lead a loving and supportive family life, and more. Smart money is a result of wealthy people applying the wisdom of a balanced life in a sensible way.

Success in the cryptosphere is more than just making money. It's about making money steadily and efficiently, with honesty and humility. This in turn leads to a greater sense of fulfillment and freedom. Becoming smart money is a lifelong process that refines our character and benefits not only ourselves, but also many generations to come. In the next four sections, we will take an in-depth exploration into smart money's systematic approach to crypto investing.

Part 2
The Crypto Investor's Cornerstone

Chapter 9

DYOR

You will often hear the phrase "**d**o **y**our **o**wn **r**esearch" (DYOR) in the cryptosphere. Since there are thousands of cryptocurrencies, this is especially relevant when you start your journey. If you don't know where to start, sorting through the dizzying amount of information strewn across the internet will be frustrating, to say the least.

To spare you from stress and strain, in this chapter we're going to explore some resources that present the most useful, well-organized information. When it comes to learning about a project's key metrics and fundamentals, the best websites are coingecko.com, coinmarketcap.com, and messari.io. They all have some basic similarities, and which you use will be a matter of personal preference.

Metrics to Watch

Let's take a look at the key features of Coingecko. In Figure 8 you can see the metrics for bitcoin (**ticker**/symbol: BTC). The current price for 1 BTC is listed in the top left, and next to that is a percentage that is either green or red. If it's green, it shows the percent increase over 24 hours, while red shows a decrease. Underneath the price, you can see the current market cap. Rankings are determined by the size of the market cap. You can also see 24 hour trading volume, 24 hour highs and lows, circulating supply, and maximum supply.

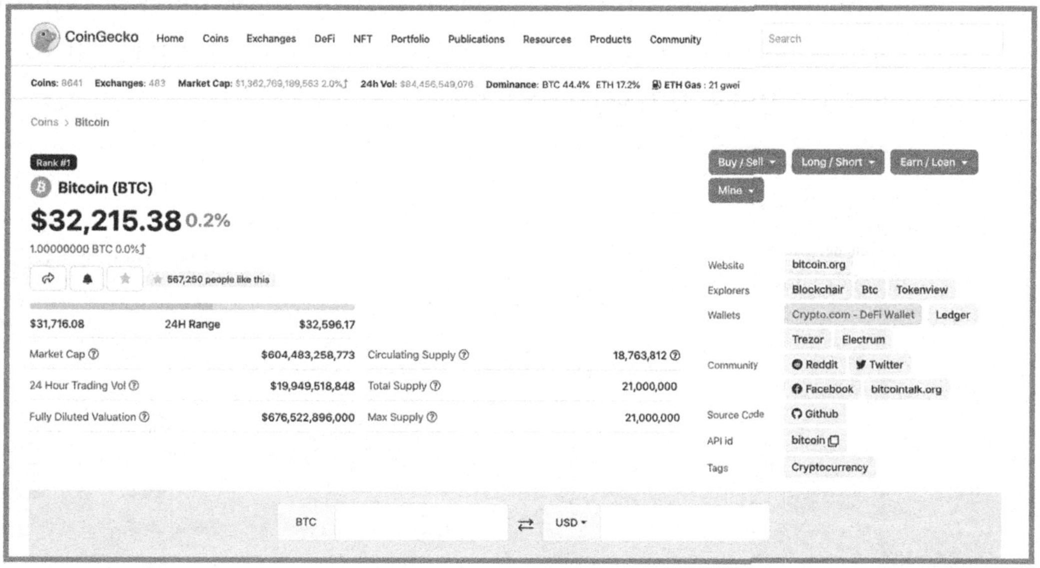

Figure 8: Bitcoin on CoinGecko.

On the right there is a link to its official website where you can read its **whitepaper**, a document that explains the purpose of a project and the problem(s) it proposes to solve. Many whitepapers have a **TL;DR**: an acronym for **t**oo **l**ong, **d**idn't **r**ead that contains a summary of the information presented.

Beneath the website, you can see its **blockchain explorers**, which are search engines for data stored on the blockchain. You can use them to track the status of transactions. The social media links are useful if you want to gauge the level of community engagement on various platforms, and the Github link is helpful if you want to track the most recent changes to code by developers. If you don't see any recent changes on Github, stay away!

If you scroll down the page, you'll see a line chart where you can track the price over many different timeframes (Figure 9). On the right you will see its high and low price over the past 1 day, 7 days, and all time. Another important metric in this section is the **market cap dominance**. This takes the market cap of the project and compares it to the market cap for the entire cryptosphere.

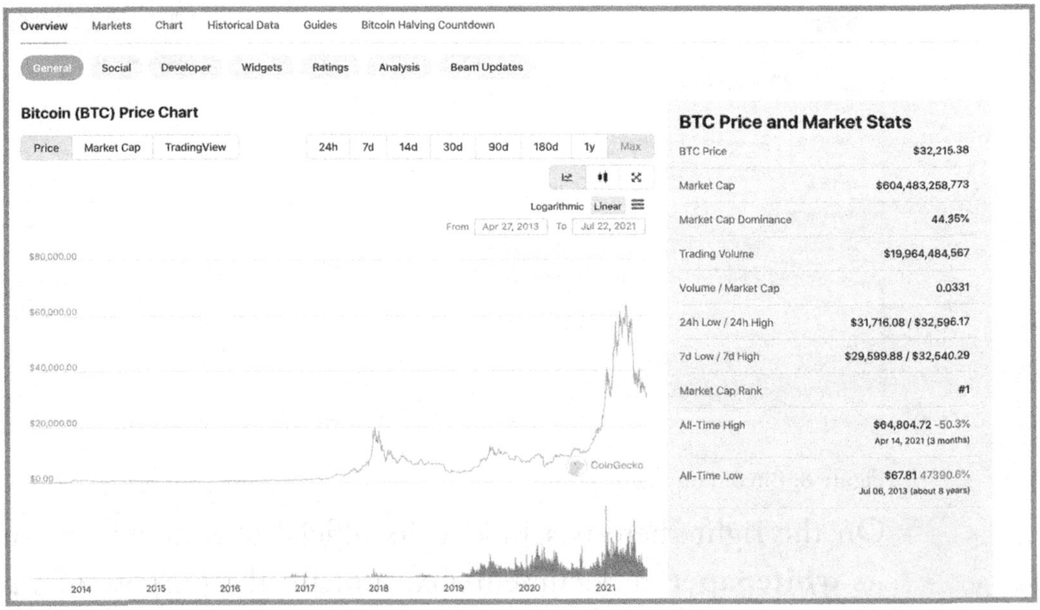

Figure 9: Scroll down to see more detailed metrics.

Continue scrolling down and click "See All Trading Pairs," which will show you which exchanges support the coin, their daily trading volume, and more (Figure 10). Ideally you are trading on exchanges with a high trust score (green), a low spread, and a minimal difference between the +2% and -2% Depth. Low spreads mean fairer buy and sell prices, while less difference in depth means more available liquidity on the exchange. We will explore these concepts in more detail later in this section.

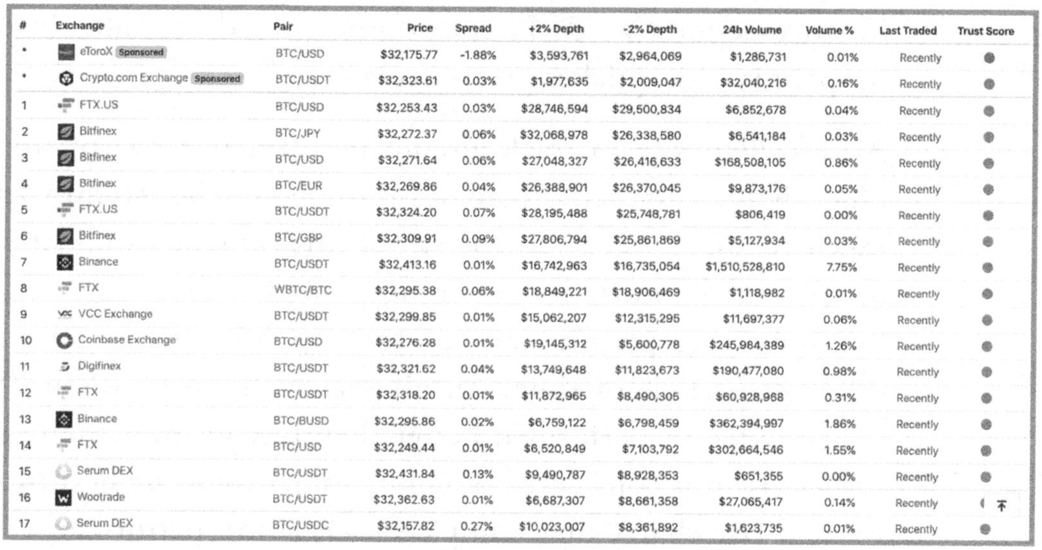

#	Exchange	Pair	Price	Spread	+2% Depth	-2% Depth	24h Volume	Volume %	Last Traded	Trust Score
•	eToroX Sponsored	BTC/USD	$32,175.77	-1.88%	$3,593,761	$2,964,069	$1,286,731	0.01%	Recently	●
•	Crypto.com Exchange Sponsored	BTC/USDT	$32,323.61	0.03%	$1,977,635	$2,009,047	$32,040,216	0.16%	Recently	●
1	FTX.US	BTC/USD	$32,253.43	0.03%	$28,746,594	$29,500,834	$6,852,678	0.04%	Recently	●
2	Bitfinex	BTC/JPY	$32,272.37	0.06%	$32,068,978	$26,338,580	$6,541,184	0.03%	Recently	●
3	Bitfinex	BTC/USD	$32,271.64	0.06%	$27,048,327	$26,416,633	$168,508,105	0.86%	Recently	●
4	Bitfinex	BTC/EUR	$32,269.86	0.04%	$26,388,901	$26,370,045	$9,873,176	0.05%	Recently	●
5	FTX.US	BTC/USDT	$32,324.20	0.07%	$28,195,488	$25,748,781	$806,419	0.00%	Recently	●
6	Bitfinex	BTC/GBP	$32,309.91	0.09%	$27,806,794	$25,861,869	$5,127,934	0.03%	Recently	●
7	Binance	BTC/USDT	$32,413.16	0.01%	$16,742,963	$16,735,054	$1,510,528,810	7.75%	Recently	●
8	FTX	WBTC/BTC	$32,295.38	0.06%	$18,849,221	$18,906,469	$1,118,982	0.01%	Recently	●
9	VCC Exchange	BTC/USDT	$32,299.85	0.01%	$15,062,207	$12,315,295	$11,697,377	0.06%	Recently	●
10	Coinbase Exchange	BTC/USD	$32,276.28	0.01%	$19,145,312	$5,600,778	$245,984,389	1.26%	Recently	●
11	Digifinex	BTC/USDT	$32,321.62	0.04%	$13,749,648	$11,823,673	$190,477,080	0.98%	Recently	●
12	FTX	BTC/USDT	$32,318.20	0.01%	$11,872,965	$8,490,305	$60,928,968	0.31%	Recently	●
13	Binance	BTC/BUSD	$32,295.86	0.02%	$6,759,122	$6,798,459	$362,394,997	1.86%	Recently	●
14	FTX	BTC/USD	$32,249.44	0.01%	$6,520,849	$7,103,792	$302,664,546	1.55%	Recently	●
15	Serum DEX	BTC/USDT	$32,431.84	0.13%	$9,490,787	$8,928,353	$651,355	0.00%	Recently	●
16	Wootrade	BTC/USDT	$32,362.63	0.01%	$6,687,307	$8,661,358	$27,065,417	0.14%	Recently	●
17	Serum DEX	BTC/USDC	$32,157.82	0.27%	$10,023,007	$8,361,892	$1,623,735	0.01%	Recently	●

Figure 10: Scroll down to the bottom of the page to see exchange data.

Why is all of this information important? Well for starters, you want to make sure that the project you are considering investing in has A. a sizeable market cap, is B. backed by a significant community, C. is consistently under development, D. is offered on reputable exchanges with a high trust score, and E. has significant trading volume.

An **altcoin** (alt) is any alternative cryptocurrency to bitcoin. Cryptocurrency **coins** represent standalone blockchain networks such as Ethereum, Avalanche, and Solana. Cryptocurrency **tokens** are issued by projects that operate using protocols from existing blockchain networks. AAVE, for instance, is an **ERC-20** token, since it follows specific protocols on Ethereum's network. Tokens are therefore a certain type of coin. To keep things simple in this book, the terms coin and token will be used interchangeably.

For the past twelve years, bitcoin has had a major influence over every other cryptocurrency. Typically the higher the **bitcoin dominance**–it's percentage of total crypto market share–the more altcoins lag behind in price action. When bitcoin dominance drops, altcoins will often outpace it with quicker price increases.

An excellent site to track altcoin performance relative to bitcoin is blockchaincenter.net/altcoin-season-index/. The **Altcoin Index** is an excellent place to search for strong investment candidates, and size up potential returns on your investment. The score is determined by comparing performance of the top 50 altcoins to bitcoin over different periods of time. An **alt-season** happens when 75% of the top 50 altcoins performed better than bitcoin over the past 90 days. Otherwise it is considered bitcoin season. There are beautiful and easy to read bar charts that compare each coin's performance to bitcoin over 30 days, 90 days, and 1 year.

After studying the altcoin index, use CoinGecko or CoinMarketCap to look at the top 30-50 cryptocurrencies by market cap. The larger the market cap, the less volatile the coin will be over time, since it takes more volume to move the price up or down. Keep in mind that this doesn't make them immune to volatility! With the exception of stablecoins, all coins can see extreme price movements either up or down on any given day.

Top 20 cryptos are often referred to as **blue-chip coins**, and are considered to be more fundamentally sound than their lower-ranked peers. Although there are exceptions (usually involving dogs), larger market caps generally mean more reputable projects that have withstood the test of time. Rankings change every day, and each year there are at least a handful of lesser known projects that swap places with their blue-chip counterparts. The trick is to be able to spot them before they make their grand entry.

Finding Diamonds in the Rough

In the search for the next big altcoin, many newbie investors make the mistake of throwing their money into a dozen or more **micro-cap coins**, which have market caps that are less than $100 million. They do this under the premise that they will go up 50 or 100x. But a smaller market cap doesn't necessarily translate into exponential gains. The smaller the market cap, the easier it is to move the price of the coin *in either direction*. These projects typically have the following features in common:

1. They have been around for a year or less.
2. They have no working product or network effect that is creating wider adoption.
3. Their tokens are available on limited exchanges, often with low trust scores, low trading volume, and a large spread due to low liquidity.

Number 3 is especially noteworthy, since this can create obstacles to buying/selling during times of congestion when liquidity is wiped from the exchanges. As an example, say there is a major sell off that happens, and you want to **panic sell**: close the trade immediately and cut your losses. Imagine how frustrating it would be to receive a much lower price than you anticipated. Or even worse, to be unable to sell at all! Unfortunately, this happens all the time.

If you set up a free account with either CoinGecko or CoinMarketCap, you will be able to star the projects you like and group them together into lists. For example, you could create a "Portfolio" list that includes all your current investments, and a "Watching" list that includes projects that may eventually become part of your portfolio in the future.

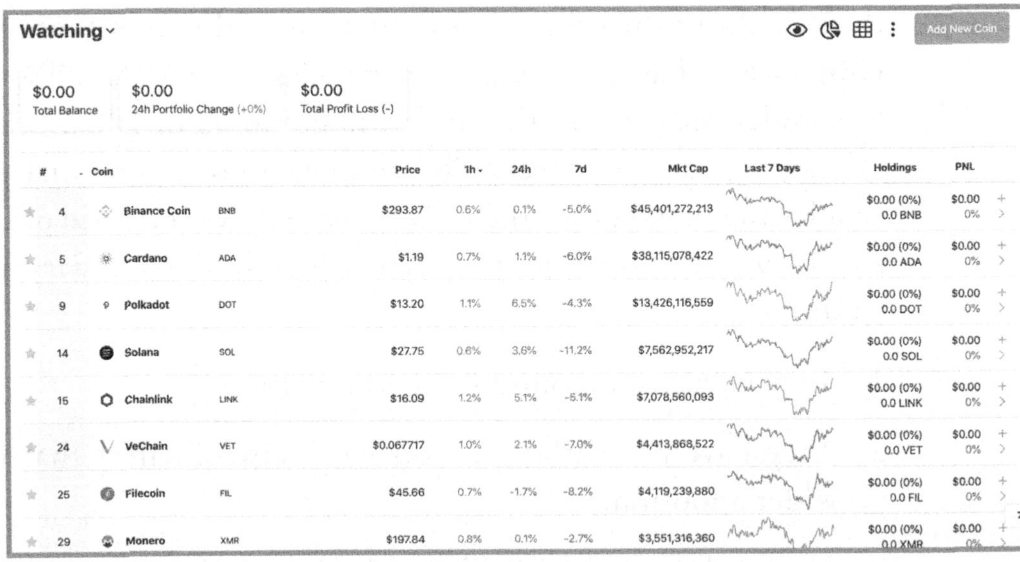

Watching

Add New Coin

$0.00 Total Balance | $0.00 24h Portfolio Change (+0%) | $0.00 Total Profit Loss (-)

#	Coin		Price	1h	24h	7d	Mkt Cap	Last 7 Days	Holdings	PNL
4	Binance Coin	BNB	$293.87	0.6%	0.1%	-5.0%	$45,401,272,213		$0.00 (0%) 0.0 BNB	$0.00 0%
5	Cardano	ADA	$1.19	0.7%	1.1%	-6.0%	$38,115,078,422		$0.00 (0%) 0.0 ADA	$0.00 0%
9	Polkadot	DOT	$13.20	1.1%	6.5%	-4.3%	$13,426,116,559		$0.00 (0%) 0.0 DOT	$0.00 0%
14	Solana	SOL	$27.75	0.6%	3.6%	-11.2%	$7,562,952,217		$0.00 (0%) 0.0 SOL	$0.00 0%
15	Chainlink	LINK	$16.09	1.2%	5.1%	-5.1%	$7,078,560,093		$0.00 (0%) 0.0 LINK	$0.00 0%
24	VeChain	VET	$0.067717	1.0%	2.1%	-7.0%	$4,413,868,522		$0.00 (0%) 0.0 VET	$0.00 0%
25	Filecoin	FIL	$45.66	0.7%	-1.7%	-8.2%	$4,119,239,880		$0.00 (0%) 0.0 FIL	$0.00 0%
29	Monero	XMR	$197.84	0.8%	0.1%	-2.7%	$3,551,316,360		$0.00 (0%) 0.0 XMR	$0.00 0%

Figure 11: Watch lists are excellent tools for side-by-side coin comparisons. Notice the "Last 7 Days" column, which is especially useful for spotting signs of strengths or weakness during significant market shifts.

The number of other people who have also starred a project can clue you in to its popularity. Sometimes lower ranked coins can have tens of thousands of followers. Verasity (VRA), for instance, has more than 4x the amount of followers than its neighbors with similar rankings.

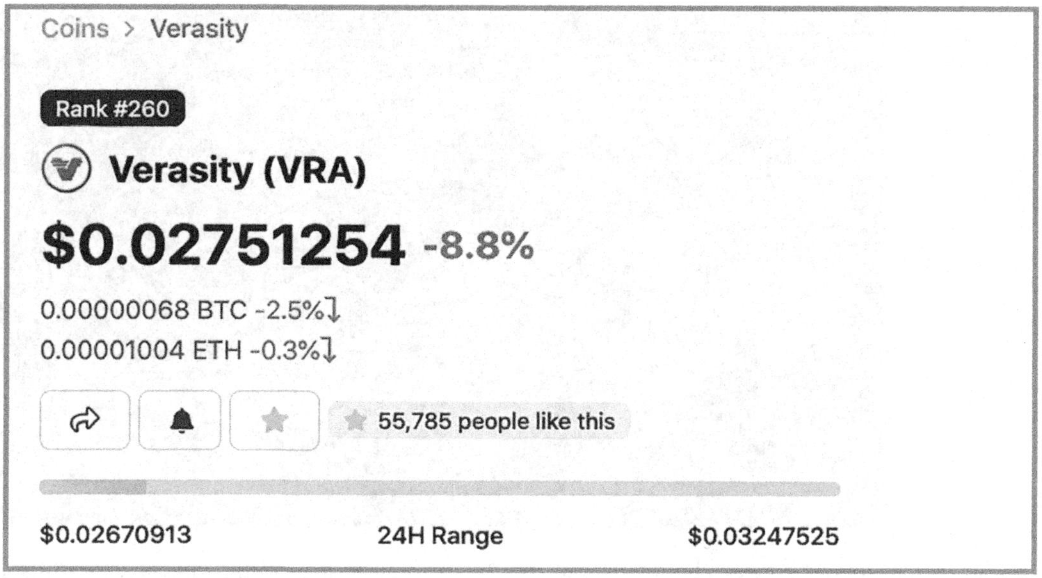

Figure 12: Verasity has 55,785 followers at the time of writing, which is extremely high relative to its #260 ranking.

Once you have created your personalized lists, you will be able to view performance comparisons of dozens of coins side-by-side. You can compare by ranking, market cap, and performance over 24 hours, 7 days, and more. The line charts on the right are helpful to get a snapshot performance comparison during significant moves either up or down. In a downtrend, for instance, you can take note of which coins are decreasing the least, and therefore showing the most strength (vice versa for an uptrend).

Cryptobubbles.net is another excellent tool for tracking crypto prices. Coins with the largest percentage gains for the chosen time period–either hour, day, week, or month–take up the most space in the display. It has an appealing layout, and makes it simple to hunt down the biggest movers. Clicking the bubble also brings up an attractive chart that shows recent price data.

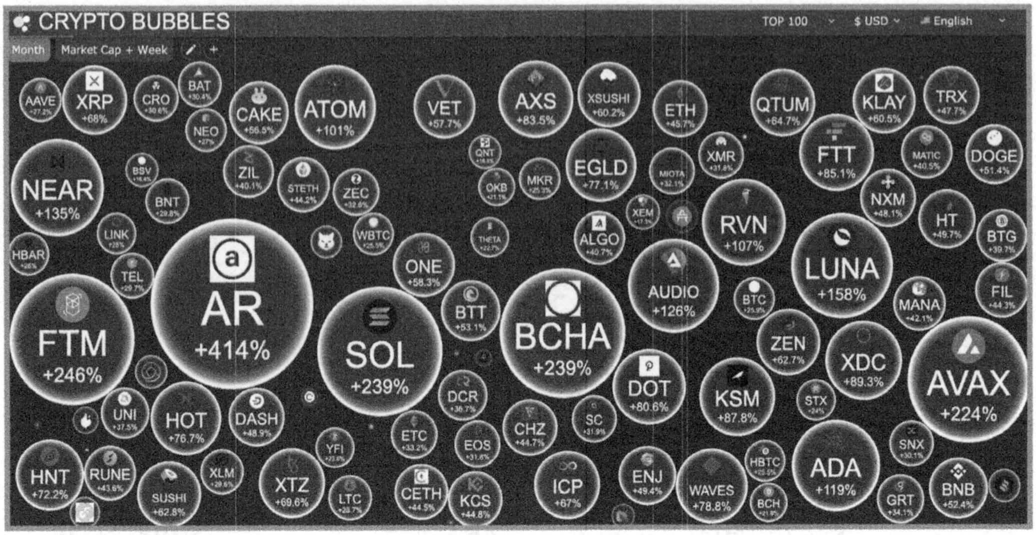

Figure 13: Cryptobubbles.net is great for price performance comparisons over different timeframes.

Chapter 10

Fundamental Analysis

Reading charts will give you clear data for a coin's past performance, but an appealing chart in the short term for a project without strong fundamentals is the recipe for disaster. **Fundamental analysis** is the process of determining the underlying strength of a project. Strong crypto projects have the following traits in common:

1. They have a real-world use case that solves (or will very soon solve) significant problems in the most efficient way possible. This is perhaps the most important factor that will determine the ultimate success of a project.

2. They have a strong team of developers with many years of experience.

3. They have a clear roadmap that includes a timeline for network upgrades and releases, and they execute on their plans.

4. They have a significant social media following and an effective marketing campaign, with regular publications to various platforms that keep investors up to speed with developments.

5. They operate using sound tokenomic principles. **Tokenomics** are the details of a coin's creation, management, and distribution.

On-Chain Analysis

On-chain data comes from activity recorded on a project's public blockchain. Since public blockchains can't be tampered with, this means that on-chain metrics can be reliable indicators of underlying strength.

Total Value Locked is the amount of money locked into the smart contracts of a blockchain network. AAVE, for instance, has over $15 billion dollars locked into its lending protocols. The larger the TVL, the more trust users have in the platform. Investors are incentivized to lock their tokens into a smart contract when there is the possibility of earning a high yield from either staking, lending, or providing liquidity (covered in Part 4). Projects with the highest TVL often pay out the best rewards, and have sleek, user-friendly interfaces.

Defillama.com provides the simplest way to compare TVL across various DeFi projects. You can also compare the Market cap/TVL ratio. The lower the ratio, the more users are locking up their holdings.

Another important on-chain metric to observe is **active wallets**: the number of wallets with a crypto balance. Active wallets are also an indication of network growth. **Metcalfe's Law** says that the value of a project is proportional to its user base squared. In other words, more connections within a network inevitably leads to a higher market cap. Ethereum is an excellent example of Metcalfe's law. The more Dapps are built on the Ethereum network, the more valuable it becomes as a currency.

Exchange inflow and outflow (to wallets) is a way to use on-chain metrics to determine investor conviction. When there is a net exchange outflow, holders are either sending their coins to wallets for storage, or locking them in a network through staking, lending, etc. Positive outflow is bullish since there is more holding than selling. Increases in exchange inflow is generally a sign that holders are preparing to sell.

You can see data for active wallets and exchange inflow/outflow on cryptoquant.com, glassnode.com, and whalemap.io. They are all worth checking out, since each has its own unique metrics. Whalemap, for example, is interesting because it specifically tracks whale inflows and outflows from 1,000 to 10,000 BTC. Glassnode deserves honorable mention for publishing a free, in-depth weekly newsletter that summarizes on-chain data so you don't have to spend hours studying their charts.

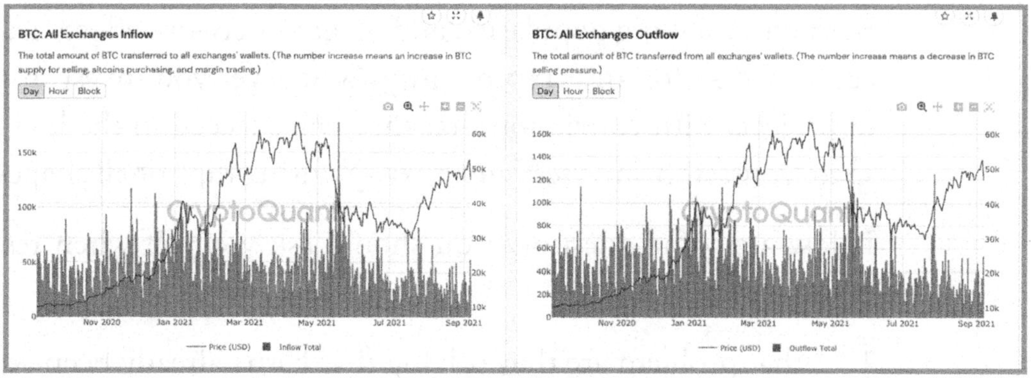

Figure 14: Exchange inflow vs. outflow on cryptoquant.com.

Crypto Entities

Have you ever noticed that businesses and organizations are often referred to as entities? This is because an organization is a living being whose vibration is determined by the collective thoughts, words, and actions of its participants.

This basic law also applies to the cryptosphere. When developers are focused on maximizing the value they provide, they don't have time to create useless hype or trash talk other projects in their niche. Such projects naturally attract large social media followings that will support them through thick and thin.

If you track high integrity projects long enough, you will notice that they don't crash as hard as the others during systematic corrections, because they have a loyal investor base. They also typically have a higher percentage of their circulating supply staked on the network by their investors. Finally, frequent developments and partnerships are common, which creates a strong **network effect**: an ever-increasing number of projects that are eager and willing to use its protocols.

Resilience, staking popularity, and obvious network effects are a winning combination for any crypto entity. Whenever you invest in these types of projects with conviction that they will succeed in the long term, you spare yourself from stress and anxiety during dips of all shapes and sizes!

Below are some questions you should ask yourself when researching a crypto entity:

1. What problem are they solving that hasn't already been addressed by others? Does their whitepaper stand out to you?

2. Is their technology both relevant and useful?
3. What is their native coin or token used for, and is it absolutely necessary?
4. Have they followed through on their plans and promises so far?
5. Have they established fair tokenomic principles that will benefit you as an investor?
6. Are you impressed by their website and social media presence?
7. What is your sense about their team of developers? Are they competent and experienced? Have they worked on other projects?
8. Can you see other projects in the future benefitting from the technology they have to offer?

Chapter 11

Tokenomics Simplified

Not all coins are created equal! The core team behind a project creates the rules for how tokens are **minted**, or created, so that they can be circulated both into and out of the network. Many projects will keep some tokens in reserve to be released into circulation gradually over time. This is called a **vesting schedule**, and helps to promote steady growth of the project.

Let's look at the tokenomics for bitcoin. Bitcoin was created using a **fair launch**, with Satoshi Nakamoto mining the first block in January 2009. A fair launch means that coins are mined gradually over time, with none allocated to investors in advance. As mentioned, bitcoin has a hard cap of 21 million coins, which makes it a deflationary asset that is one of its hallmark features as a store of value.

Nakamoto also designed the protocol so that there would be a slow, steady stream of coins that were minted as block rewards for validating blocks of transactions. There are mining difficulty adjustments every two weeks that ensure that it takes roughly 10 minutes to mine a block of transactions. There are also **halving** (aka halvening) events where the block reward is cut in half approximately every four years (after 210,000 bitcoin blocks have been mined). During the first four years after the genesis block was mined, the reward was 50 BTC per block. As of May 11, 2020, miners receive 6.25 BTC per block. The inevitable supply shock created after each halving has been one cause of the exponential growth in bitcoin's price over the years.

ICOs, IEOs, and IDOs

Nowadays, most crypto start-ups opt out of fair launches. Back in 2017, many began to hold **Initial Coin Offerings**, or ICOs, to raise funds directly on their website for the project before launch. Basically, they sold a percentage of pre-mined coins to early investors and venture capitalists (VCs), who often received a major discount for investing early. These projects would also allocate a portion of their supply to founders and team members.

As an example, consider Ethereum. Most of its initial circulating supply was pre-mined to be offered to early investors during their ICO in 2014. Ethereum has no hard cap to allow for continued growth of token supply as the network expands and more transactions are made.

There have been several issues with ICOs that have caused them to fall out of favor in recent years. For starters, there were several **exit scams**, where project founders would simply take the money raised and run. This drew undesirable attention from the SEC, which labeled ICOs as unregistered securities sales.

After China banned ICOs in 2017, crypto projects began resorting to **Initial Exchange Offerings** (IEO) instead. IEOs are fundraisers that happen directly on exchanges, which have the benefit of increased trust and transparency. Binance Launchpad was the first IEO platform, which charges crypto start-ups a fee to fundraise on the exchange. Tokens are listed on the exchange after completion of the IEO, enhancing a project's credibility in the eyes of investors.

The downsides to IEOs are that exchanges can charge high fees to crypto start-ups, and also prohibit them from offering their coins on other exchanges. As a result, some crypto start-ups are now opting to use **Initial Dex Offerings** (IDO). In this case, the project launches its token on a decentralized exchange to raise funding directly from retail investors. While the ability to immediately trade the tokens makes them similar to IEOs, what makes IDOs more attractive to start-ups is their lower listing costs.

Who Owns What?

Regardless of how a project raised its initial funds, it is helpful to investigate how the current circulating supply of a token is distributed. Some blockchain explorers such as etherscan.com will allow you to view the top 1,000 holders of a project, and their percentage holdings. In some cases, this information can be very revealing. For example, if you look at Dogecoin on blockchair.com, you will notice that one wallet holds about 30% of the entire circulating supply!

You can also learn more about tokenomics for a project at messari.io (click Profile, then Token Economics). Examples of red flags would be any pieces of the pie that are too heavily weighted toward VCs and/

or team members (aka foundation), and are distributed all at once as opposed to gradually over time. In these cases there is nothing stopping these whales from dumping their supply at your expense!

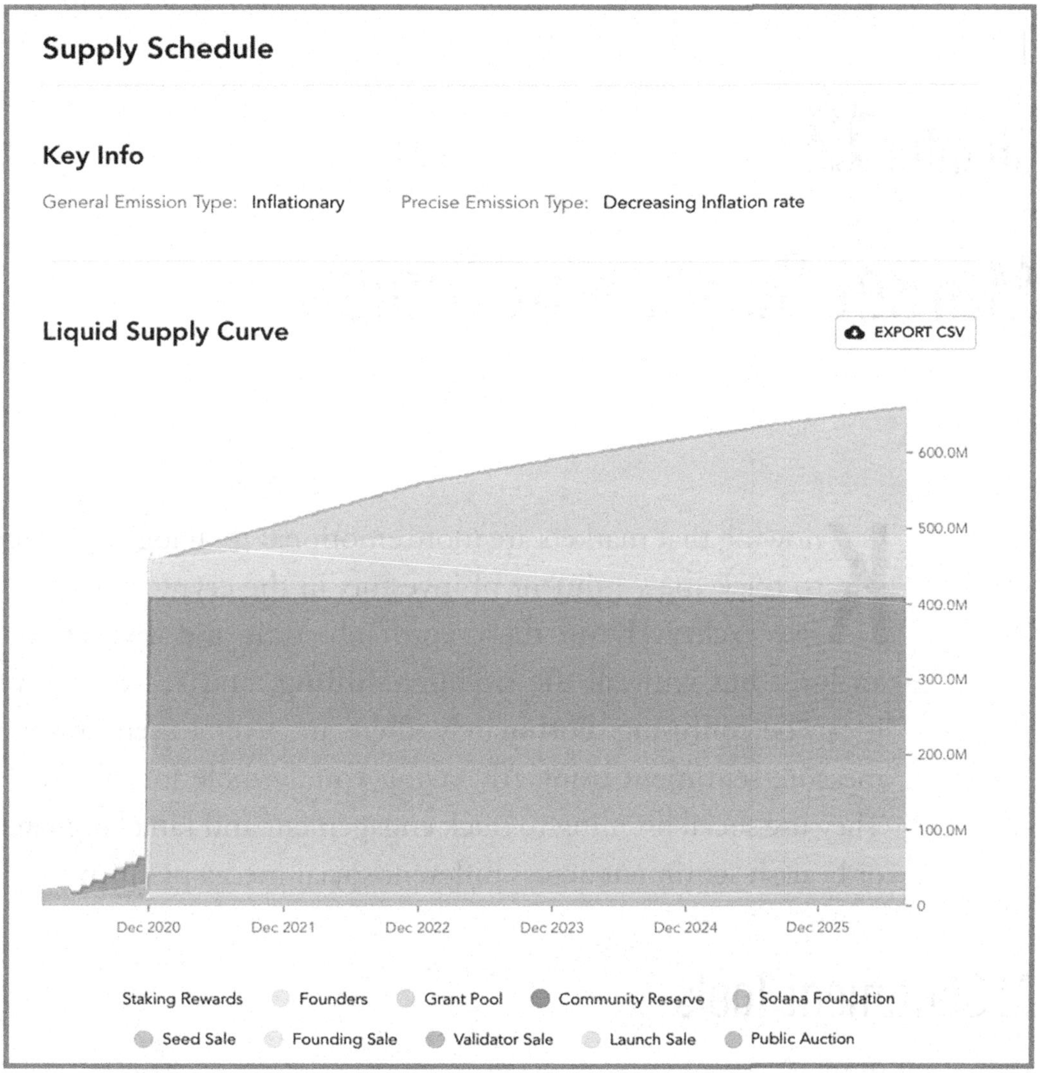

Figure 15: SOL's liquid supply curve on messari.io, also known as a vesting schedule. Notice the steep incline in January of 2021, which reveals a large amount of coins quickly entering circulating supply. A more gradual vesting schedule would be less of a concern, since there is a lower likelihood that VCs and team members could have as drastic an impact on the price should they choose to sell their bags.

Chapter **12**

Market Sentiment Tools

Knowing that markets are more emotional than logical is reason enough to track the sentiment of investors in the cryptosphere. Some people rely exclusively on the crypto subreddit and Twitter for sentiment analysis, but with all the trolling, **shilling**, and self-proclaimed experts, it can be confusing. Fortunately, there are several useful online tools that measure sentiment using AI, saving you a whole lot of time and energy. They use social listening to track engagement and rank language for bullish or bearish sentiment across different social media platforms.

AI Sentiment Tools

1. The Crypto Fear & Greed Index

The Crypto Fear & Greed Index at alternative.me/crypto/ is one of the most popular sentiment analysis tools in crypto (Figure 16). Its sole focus is bitcoin, since it carries the most weight in the cryptosphere. It combines volatility, momentum, social media interactions, bitcoin dominance, and google search trends to give a numerical ranking on a scale of 0-100 between extreme fear and extreme greed. Beneath the ranking you can view historical data on a line chart for convenient comparison. The index is the crypto version of the original fear and greed index created by CNN as a public sentiment analysis tool for the stock market.

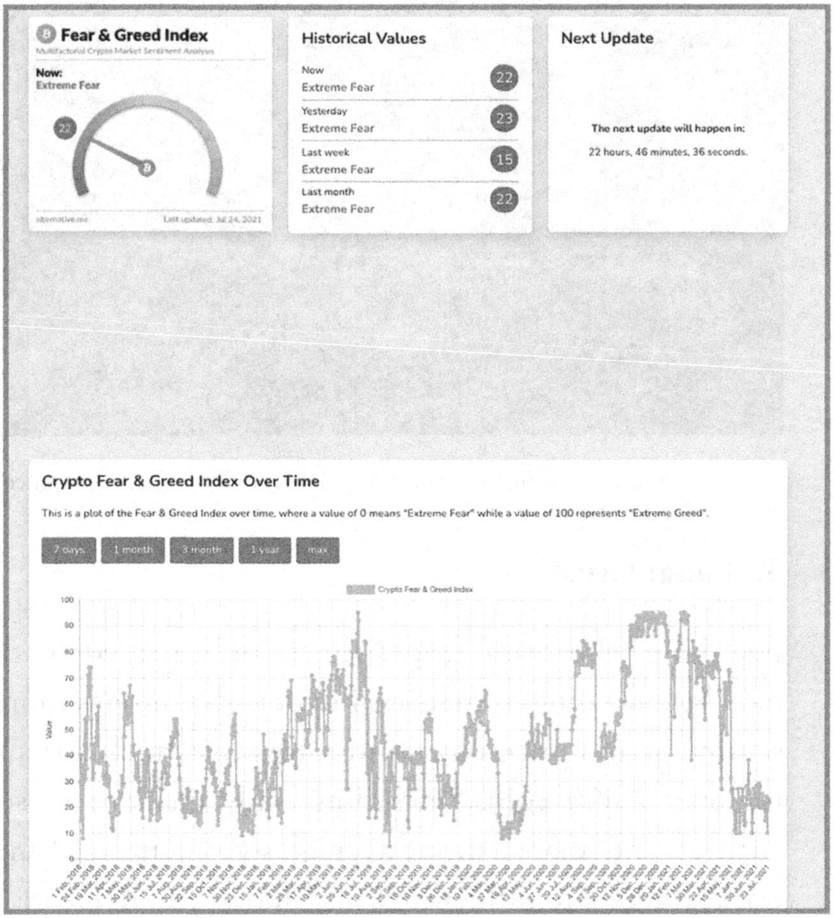

Figure 16: The Crypto Fear & Greed Index. Extreme fear means it's not a fun day for crypto!

2. The Bitcoin Bull & Bear Index

The Bitcoin Bull & Bear Index at augmento.ai/bitcoin-sentiment/ is another excellent sentiment analysis tool that uses AI to track social media Bitcoin conversations on Twitter, Reddit, and Bitcointalk. A ranking of 1 is extreme bullish, while 0 is extreme bearish. As you can see from their chart, there is sometimes, but not always a correlation between sentiment and the price of bitcoin.

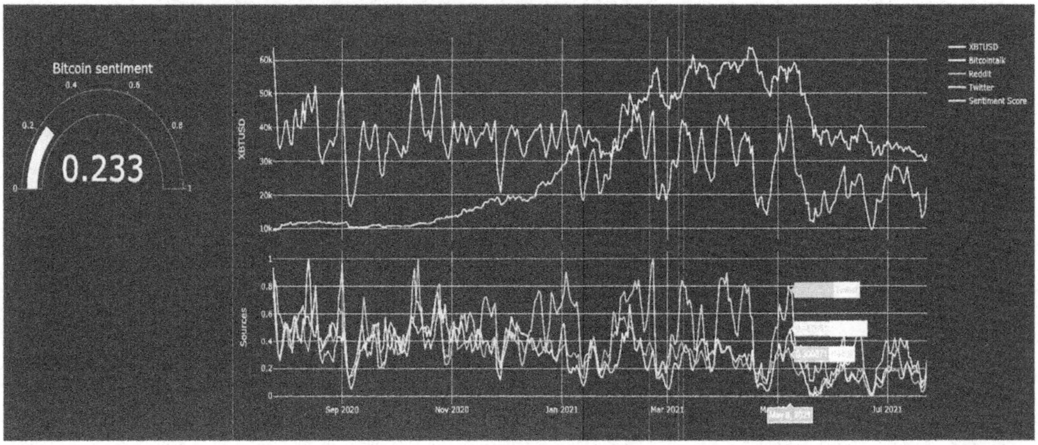

Figure 17: Augmento.ai bitcoin sentiment analysis tool is an aggregate of conversation sentiment on Twitter, Reddit, and Bitcointalk.

3. Lunar Crush

For those who want to get more sophisticated in their analysis, and look beyond Bitcoin metrics, lunarcrush.com has more than twenty five sentiment and price-based indicators that track dozens of social metrics for over 2,000 coins, including daily changes in social volume and engagement, shared links, average sentiment, and much more. It also tracks top crypto influencers, and allows you to create a favorites list and side-by-side comparisons of coins.

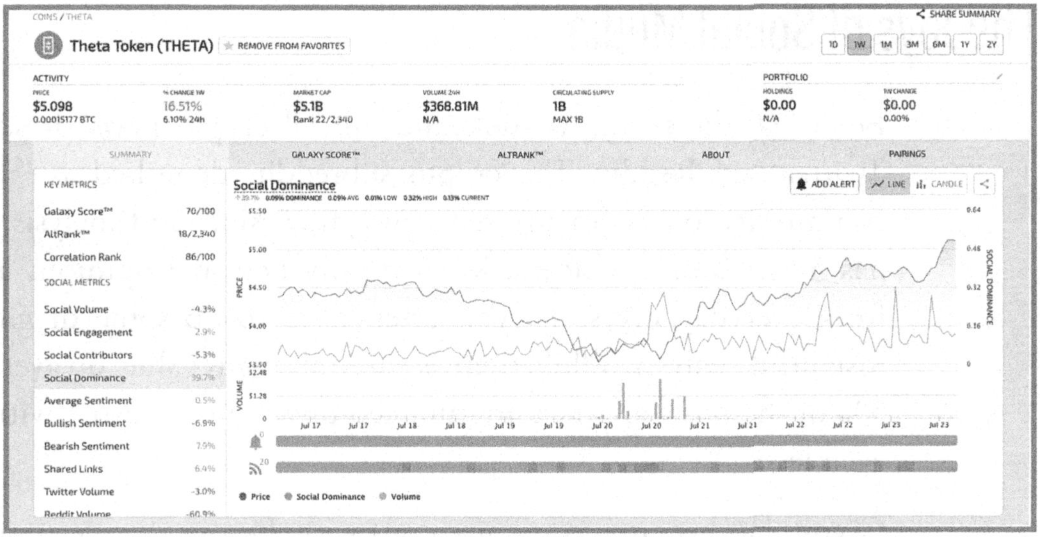

Figure 18: Lunar Crush offers in-depth sentiment analysis for over 2,000 coins.

Crypto Media Sources

Following various media sources is another way to observe market sentiment. While it isn't nearly as objective as AI based sentiment analysis, sometimes a quick scroll through the likes of Cointelegraph, Coindesk, Decrypt, or CryptoPotato can reveal clearer reasons why specific coins are either pumping or dumping. If you aren't a fan of reading the news, check out CryptoPanic, which is a news aggregator that gives one line summaries of market developments every hour of the day.

Positive sentiment for a project could be an announcement about a partnership, a major network upgrade, or receiving capital from a well known institution. Negative sentiment is anything that compromises the project in any way, whether it be a delayed upgrade, unexpected network issues or hacks, or announcements about possible regulatory actions.

The Joys of Social Media

Before all the fancy AI tools and official crypto news sites, there was Twitter and Reddit. The crypto subreddit, described as "the leading community for cryptocurrency news, discussion, and analysis," currently has 3.4 million members. With mostly personal opinions and stories from everyday investors, you often need to do some digging to find insightful analysis. While there is entertainment value, on average maybe 5% of the posts are meaningful deep dives into crypto. Twitter may be even less.

Crypto has become so synonymous with Twitter that cryptonauts have created their own imaginary fork: **crypto twitter** (CT). To the untrained eye, CT looks exactly the same as the original version. In fact, it *is* exactly the same, aside from the fact that crypto fanatics on Twitter like to use fancy lingo, and often exhibit borderline psychotic passive aggressive tendencies that bewilder and alienate nocoiners and **newbs**/normies (both slang for newer, less experienced investors).

In truth, Twitter isn't really all that bad, especially with Bitcoin maxi Jack Dorsey at the helm. To be fair, you'll find some crazies in every arena of life, and there are plenty of well-meaning crypto veterans who are keeping the balance and educating the masses. Still, it takes some patience and discernment to filter out all the noise.

If you aren't already entrenched in the game, the best place to get started is to find influencers you admire and see who they follow. You will know when someone is worth following when you feel like taking notes on their posts and threads (check out DeFiMagick). Instead of scribbling pearls of wisdom on bits of paper, however, you should sign up for readwise.io, which allows you to save threads to refer back to for later reading.

Price Trends and Alt-seasons

Even with all the detailed metrics, market sentiment isn't easy to measure, and there are many other factors that affect price. It will mainly depend on whether crypto is facing a bull or bear market. A **bull market** is obvious when the long term price trend (several months to years) is up, while a **bear market** is marked by a long term downtrend.

News is resolved in the direction of the overall trend. In a bull market, even bearish news can be easily digested without a significant impact on price. During a bear market, however, even news that entire countries are adopting bitcoin won't be enough to nudge the price out of its slump.

Another factor that influences market sentiment is whether bitcoin is dominant or an alt-season is underway. When an alt-season is in full swing, the bitcoin blues will have much less of an impact on alts. During this time altcoins will see explosive growth, as even neutral events can lead to quick price surges. In contrast, when bitcoin is dominant, alts will take a backseat, regardless of recent developments.

Take It with Salt

The main reason you shouldn't give too much emphasis to market sentiment is because the market is primarily controlled by **whales**: investors that own significant quantities of cryptocurrency. Bitcoin whales are considered to own at least 1,000 BTC. Some whales are institutions including governments, corporations, and large financial firms. Others are high net-worth individuals, many of whom happened to invest in crypto while it was still relatively unknown.

On-chain data suggests that 85% of cryptocurrency volume is traded by 4% of investors. While it is true that these whale investors are also paying attention to the media (and possibly manipulating it as well), it would be naive to think that they are making all of their decisions based on what people believe on CT, or CNBC's latest FUD campaign.

Moonboys and Bad News Bears

Most of what you will read on the internet is rubbish. The most popular voices are usually the loudest and most obnoxious. Unsurprisingly, they are also the best at appealing to extreme emotions in their viewers. On one side you have **moonboys**, who are shamelessly shilling their bags for their own personal gain. Then there are the **bad news bears**, who have the uncanny ability to spin crypto conspiracy theories that will leave you questioning your entire life purpose. Listening to them for even five minutes can be enough to convince you that crypto is finally about to keel over.

It is said that our closest associations will ultimately either make or break us. Imagine for a moment that a bad news bear gets you down about your investments. You hastily decide to sell everything and revert to the stress free life of a nocoiner. Lo and behold, less than a month later the market recovers, and you watch in agony as the coins you abandoned leapfrog out of their funk and into the stratosphere. Yet another crypto fairytale gone horribly wrong. Sadly, it happens during both bull and bear markets.

The good news is that there are many conscious crypto investors who are releasing incredibly informative content on a daily basis. Coin Bureau, for example, is one of the best YouTube channels out there for all things crypto. Not only does Guy provide stellar content that is thoroughly researched, he does so in a clear and organized way that gives traditional media outlets a run for their money. Guy offers updates on popular coins, deep dives into lesser known projects with promising technology, reviews of popular crypto services, weekly news summaries, and so much more.

There are of course dozens of other notable YouTube channels, including Ellio Trades Crypto, AltCoin Daily, Into the Cryptoverse, CoinsKid, and more. Whoever you choose to give your attention, what's most important is that you use your discernment when browsing through crypto-related content, and hone in on only the most trustworthy sources. When in doubt, leave them out!

Chapter 13

Crypto Categories

Remember that cryptocurrency is an internet of value, which means that it will have utility that stretches far beyond financial applications. In fact, there are so many exciting possibilities that are either already being developed, or will be over the next several decades, that we will look back and marvel in the same way we do today at the evolution of the internet.

In this section, we are going to explore ten major crypto categories in more detail. While many more categories will inevitably be created in the years to come, this list is the most comprehensive to date. As you read about each category, see which ones have the most appeal to you as an investor, as this will help you to begin to narrow down your prospects.

1. Store of Value Cryptocurrencies (SOV)

Store of value (SOV) coins are like digital gold, since they either maintain or increase purchasing power over time. Inflation makes the dollar a terrible store of value. One hundred dollars deposited into a savings account in 1913 would buy you about $2 worth of goods today. Even though we are earning much more than Americans did a century ago, our dollars have less purchasing power. Bitcoin, on the other hand, is considered a superior store of value because it is a deflationary asset with a fixed supply, and the most secure payment network on the planet.

SOV coins are generally considered to be safer investments that are less volatile over time when compared to other cryptocurrencies. Their main disadvantage is that their functionality is limited. They often receive criticism for their inability to evolve at the rate of other blockchains. While this may be true, it isn't a fair criticism since they are designed for security and stability. Bitcoin may take much longer to process transactions than some other blockchains, but this gives its network the resilience it has needed to thrive for over twelve years now.

Cryptocurrencies with a fixed supply that received a fair launch and are mined using PoW are prime candidates for the SOV title. Bitcoin and Litecoin are excellent examples. Inflationary cryptocurrencies that were pre-mined for ICO sales aren't considered stores of value.

2. Smart Contract Cryptocurrencies

Some cryptocurrencies use smart contracts that perform a function automatically when a specific set of conditions are met. Remember that smart contracts are immutable and decentralized, which means they can't be modified or shut down by a central authority. Ethereum (ETH) and Cardano (ADA) are the two largest smart contract cryptocurrencies by market cap.

As a refresher, Dapps are combinations of smart contracts used for trading, lending, borrowing, and more. Millions of daily users interact with over 5,000 Dapps hosted by over two dozen smart contract blockchains. No middle-man is necessary for Dapps to operate. You don't need to give up your personal information–all you need to use them is an internet connection.

Smart contract cryptos have the unique ability to create different types of tokens that are either fungible or non-fungible. **Fungible** tokens are identical to one another, and can easily be exchanged and divided, just like paper money. **Non-fungible tokens** (NFTs) represent unique, or collectible items that aren't interchangeable, and can't be divided. They essentially create the ability to digitize assets, and are going to revolutionize numerous industries.

NFTs are mostly minted using the Ethereum network, and are purchased in ETH as well, although this is likely to change as competing networks scale and lower minting costs. Some of the many examples of NFT use cases include digital art and other virtual assets, medical data, voting and elections, supply chain tracking, and real-world assets such as a car title or house deed. NFT P2P marketplaces such as Rarible, Opensea, and Mintable allow you to make offers on thousands of NFTs across various categories.

The value of smart contract cryptocurrencies lies in their utility as payment for network transactions. The more successful Dapps that are built on a smart contract blockchain, the greater the network effect and therefore the higher the value of its coins. Ethereum currently has the largest network effect of any smart contract cryptocurrency. As a result, many ETH maxis firmly believe in **the flippening**: the theory that ethereum's market cap will one day surpass that of bitcoin. Visit buybitcoinworldwide.com/flippening/ to see side by side comparisons of their market caps, trading volumes, and mining rewards.

3. Payment Cryptocurrencies

Payment processors such as Visa, Mastercard, and PayPal make their money in transaction fees. They use outdated networks that are slow and inefficient. Payment cryptocurrencies, on the other hand, allow for transactions to be finalized in seconds at an extremely low cost. What's more, you retain full control of your funds in your personal wallet (covered in Part 4). You don't need to rely on banks for custody, or worry about payment issues or other account restrictions. Popular payment coins include Terra (LUNA), Stellar (XLM), and Bitcoin Cash (BCH). Some payment cryptos also implement smart contracts and Dapp ecosystems to allow for ease of use.

The advantage of payment cryptos is that they have the highest chances of adoption, since there is a high demand for more efficient methods of transferring currency across borders. The current disadvantages are that payment cryptos don't yet have a stable price, and will still take some time to reach a significant level of adoption.

Keep in mind that payment cryptocurrencies aren't meant to compete directly with fiat currencies, but rather serve as an alternative mode of transfer. For example, let's say someone from Argentina who is working in the United States wants to quickly send his monthly paycheck back to his family. He can onboard his fiat, convert it to DASH, and send it through an app on his phone in a matter of minutes. His family can then either cash it out or convert it into other cryptocurrencies. By doing this, he will save an average of 7% in remittance fees, and up to a week of processing time. Recipients who worry that the coins will decrease in value can simply convert them into stablecoins, and then store them in their personal wallet. No banks = no hassle.

4. Social Tokens

Over the course of the next decade, we will likely see every major brand and social media platform create its own **social token**. Imagine owning DISN tokens, and receiving loyalty rewards that grant premiere access to new Disney titles, or monthly subscription discounts according to your investment tier. Or getting paid in ASCEND tokens for creating meaningful social media content that users upvote. Or having your KOMB public address scanned using the QR code in your crypto wallet at the health food shop to receive tokens of gratitude for buying a kombucha. You could theoretically swap them for other tokens, or redeem them for more fizzy deliciousness.

Social tokens are destined for stardom because they create a sense of inclusion by encouraging community participation. Companies will be able to leverage their massive followings by integrating social tokens into virtual hubs where fans can both interact and transact. Disney, for example, has already spent decades proving its value to society. Now it simply needs to add the infrastructure to buy and spend the DISN token.

One subcategory of social tokens that is driving massive adoption in crypto are gaming tokens. Did you know that there are an estimated 1 billion online gamers worldwide? [1] In 2020, online gaming generated over $21 billion in revenues, up 22% from 2019. During the pandemic, many people who were forced to stay home turned to gaming for both entertainment and interpersonal connection.

Thanks to blockchain technology, innovators in the online gaming world are empowering players to not only engage in thriving social communities, but to participate in virtual economies by earning cryptocurrency as well. Axie Infinity is one such game that has become a trailblazer in the **play-to-earn** (P2E) frontier. Players can enter the virtual realm of Lunacia and quest for tokens by guiding Axies, which are colorful blobs that resemble Pokémon. Sky Mavis, creator of Axie Infinity, has a very clear formula for success: combine addictive gameplay with an item economy where rewards are paid in cryptocurrencies.

Players of the game can earn Smooth Love Potions (SLP) and Axie Infinity Shards (AXS), and can then either sell them in the cryptosphere or use them to breed more Axies. The Axies are collectible NFTs that can be traded or sold for ETH in the Axie Marketplace.

Axie Infinity exploded in popularity in a few short months, increasing from 30,000 daily users in April 2021 to nearly one million daily users only four months later.[2] Its price increased over 4,500% during that time as it skyrocketed to the top 40 by market cap. What was most impressive about the run of AXS was that it coincided with a systematic correction where nearly every other altcoin dropped 60-80%!

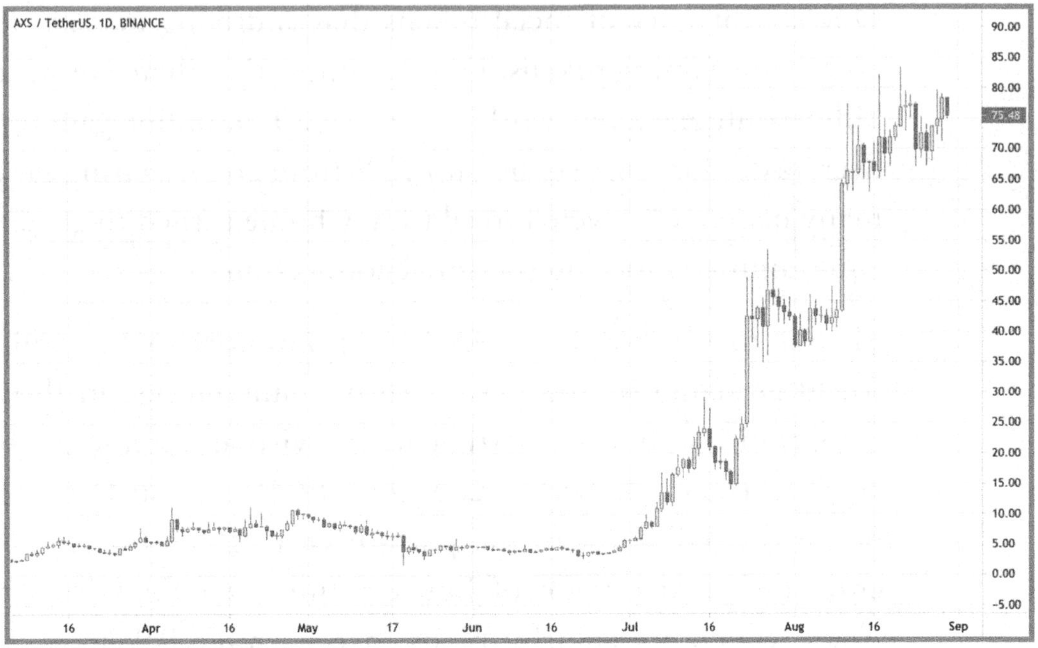

Figure 19: Axie Infinity (AXS) outperformed every other token in July/August 2021.

Even though Axie Infinity is literally child's play, it represents a far greater potential in the cryptosphere. When it comes to investing in the digital age, the most entertaining option might just be the most likely to succeed. Gaming is dominating the internet, and its popularity will only grow as blockchain technology is applied to create novel ecosystems with financial perks for players. It is definitely possible that gaming tokens will be the first to **decouple** from bitcoin. In other words, their price will no longer depend on bitcoin's price action.

There will be dozens, if not hundreds more P2E games with item economies that incentivize players to join in droves. Some notable examples of promising gaming projects are Illuvium (ILV), Star Atlas (ATLAS), and Sandbox (SAND).

5. Video Streaming Cryptocurrencies

Did you know that the demand for video content on the internet has been growing exponentially in recent years? At the current rate of growth, 82% of internet traffic will be video content by 2022.[3] Revenues of streaming companies such as HBO Max, Netflix, and Disney Plus are projected to triple by that time.

Demand for high quality content is soaring, and content delivery providers (CDPs), which are currently centralized, are struggling to keep up. The result is poorer quality video and longer load times in many parts of the world. What's more, CDPs prevent revenues from flowing back to content creators who deserve them the most.

The solution? Decentralized video streaming networks that are powered by blockchain technology, including Theta, Livepeer (LPT) and Moviebloc (MBL). Theta, for example, uses multiple layers of nodes–tens of thousands and climbing–that relay video content across the Theta network. Anyone can become an edge node by sharing their internet bandwidth and computing resources P2P.

Node expansion lowers the cost for video content distributors, and more importantly, makes it possible to stream both 4k and 8k quality video while minimizing buffering. Nodes are incentivized by earning rewards in TFuel. Viewers can also earn TFuel just by watching streams on the network. Theta also supports smart contracts, allowing for other Dapps to be built on the network that can further enhance the protocol.

With YouTube already accounting for 35% of global internet traffic, there will be a great need for decentralized video streaming to enhance content delivery in the decades to come.[4] It is quite possible that many centralized CDPs will partner with their decentralized counterparts in the near future, who can offer cheaper and more efficient video streams to the masses.

6. Privacy Coins

All blockchains can be viewed by the public, which means you can see every transaction as it takes place. Activity can be easily monitored by anyone with an internet connection using blockchain explorers. While public addresses aren't directly linked to identities, it isn't difficult for "trusted parties" such as exchanges to connect the two, considering they require identity verification at sign up.

Full transparency may not seem like a big deal at first glance, but would you want every transaction you've ever made available to prying eyes? How about your full wallet balance? As an example, if you give out your public Ethereum address to receive an ETH payment, anyone can simply plug it into Etherscan and see token balances of any and all ERC-20 tokens you own, including a complete history of all the transactions you've ever made! Try it sometime, and see for yourself.

Privacy coins address these concerns from several different angles. Tornado Cash (TORN), for instance, mixes up your ETH funds and sends them to a new address with no transaction history. The Secret network (SCRT) is focused on developing privacy Dapps. Monero (XMR) and Zcash (ZEC) privatize transactions using sophisticated technology that has continued to be immune to government hackers across the globe. In fact, the IRS has a $625,000 bounty on anyone who can crack Monero's code! [5]

The main disadvantage of privacy coins is that many fear that they may become targets of regulators. As a result, privacy coins have at times been delisted from several different exchanges, which can adversely affect the price. Still, there will always be decentralized exchanges and Dapps that allow for them to be swapped free from the clutches of centralized authorities. As long as anonymity is a concern, privacy coins will continue to evolve and flourish.

7. Data Storage Cryptocurrencies

We are all familiar with existing centralized storage solutions, including Amazon Web Services, Google Cloud, and ICloud. The problem is that they are private servers, which can monopolize prices and limit accessibility. This creates restrictive barriers to billions of people, many of whom need to access data from remote areas. What's more, their centralization means they are susceptible to hacks and other security breaches. Data storage cryptocurrencies aim to solve this problem by using a P2P file storage system. Examples include Filecoin (FIL), Bitorrent (BTT), and Arweave (AR).

Filecoin, for example, uses a special protocol where each file that is uploaded to the network is assigned a unique signature. Multiple miners across the globe will have the option to host the file, and can bid to win storage contracts from network users. Users who wish to access their files can do so at any time by searching for miners who host their file(s) and then choosing the most affordable option. FIL coins are used as the currency for users to pay miners each time they want to access their files.

Since miners anywhere in the world can store a copy of the file, it creates a secure global network. The more popular a file becomes, the more miners host it, which makes the file both cheaper and more accessible. Miners who validate blocks and relay data are kept honest since they must prove to the network that they are storing files using cryptographic proofs. If they are unable to provide proof, they face **slashing penalties**: they must forfeit a portion of the FIL coins they have put up as collateral to serve as miners.

With a growing demand for data storage by individuals, businesses, and institutions, it is only a matter of time before decentralized storage begins to gain more traction.

8. Oracle Cryptocurrencies

Oracle coins bring real world data to smart contract blockchains. Most of the applications we use every day use data feeds. Some examples include weather info., the date and time, and item prices on various websites. Centralized applications relay this information using advanced programming interfaces (APIs) which are provided to them by another entity.

What makes oracle cryptos unique is that they are decentralized. Oracle coins are used to pay fees to retrieve data on the blockchain. There are about a dozen oracle cryptos, with Chainlink (LINK) being the most well known. The decentralized lending protocol AAVE uses LINK oracles to display real-time value of collateralized assets. Other examples include Band Protocol (BAND), API3, and Tellor (TRB). Oracles are a vital aspect of the cryptosphere, and their demand will continue to grow alongside the number of Dapps and their user base.

The main disadvantage of oracle cryptos is that a large majority of them have sub-par tokenomics, with a significant amount of their pre-mined token supply allocated to teams and private investors. As the price rises, it is likely that this will adversely affect the price. Something else to consider is that smart contract cryptos tend to use multiple different oracles for their data, which also dilutes their value. Still, oracles help to bridge the gap between the digital world and the cryptosphere, and are therefore a smart choice in a diversified portfolio.

9. Exchange Cryptocurrencies

Crypto exchanges such as FTX (FTT), Kucoin (KCS), and Binance (BNB) issue tokens to raise additional funds for their operations, as well as to fund other projects. Exchange tokens differ from company stocks in that they come with perks that are comparable to a membership subscription. For example, holding specific token quantities can earn you higher trading fee discounts, enable you to use crypto credit cards with lower fees, and grant VIP access to early token sales through IEOs.

Some exchanges will regularly perform a **buyback-and-burn**, where they buy back a portion of the circulating supply and then remove it from circulation. This is an irreversible transaction that can be viewed on a blockchain explorer, and increases the price of existing tokens by reducing the total supply. Binance, for example, takes 20% of its profits from trading fees each quarter to buyback-and-burn BNB coins. BNB burns are scheduled to take place every quarter until the total circulating supply is cut in half. This creates a deflationary effect, since BNB has a fixed supply. Binance coins have additional utility as a smart contract cryptocurrency, since the Binance Smart Chain requires the use of BNB to pay for gas fees on its network.

The main advantage of exchange tokens is that they are likely to go up in price as adoption spreads. They incentivize their users with significant rewards that benefit every type of investor. The main disadvantage is that negative press created by regulatory concerns tends to have a more drastic effect on their prices compared to other cryptocurrencies.

10. Meme Coins

Meme coins take their rightful place last in line to grab a piece of your portfolio pie. There are countless meme coins in the cryptosphere, which are typically recognized by a viral image plastered everywhere by social media fanatics. Crypto veterans can recognize worthless coins when they see them, and steer clear. Unfortunately, the average retail investor cannot. *Meme coins are the most manipulated coins in the market.* They are characterized by hype and a get-rich-quick vibe. They don't have any real world utility, and the endgame for the developers is simple: take the money and run.

Memes coins usually rally around funny jokes or attention-grabbing sentiments, and aren't limited to dog images (there are dozens of dog meme coins). Safemoon, for example, can be considered a meme coin since its "to the moon" slogan reflects the profiteering mindset of the casual investor. It punishes sellers, who lose 10% of their holdings, and then claims to reward holders with an additional 5% of tokens. But why would a reputable cryptocurrency need to punish sellers in the first place?

When the average person sees the price of Safemoon at $0.0000033 per token, they often assume that it will be easier to drive the price up higher. You could theoretically invest $100 and receive over 30 million tokens. Many fall victim to the numbers illusion, not realizing that this is a drop in the bucket when compared to a circulating supply of almost 600 trillion.

DOGE has become an interesting outlier in the meme coin category, for two main reasons. First, Elon Musk has decided to be its official spokesperson. This is significant, only because his stature as one of the wealthiest, most powerful people in the world does in fact give him the power to sway the collective into adopting a random cryptocurrency as a method of payment.[6] The second reason is that it can be merge-mined with Litecoin, which means there is a greater possibility of future development and utility. Much to the chagrin of many old-timers, it isn't likely to go away any time soon.

For the conscious crypto investor, the risks of investing in meme coins far outweigh the rewards. Where there is market manipulation, there are rapid and unpredictable price swings that reveal symptoms of chronic instability. While some meme coins may withstand the test of time, until they achieve actual network effects, it is wise to keep your distance.

Making the Choice

It is clearly impossible to invest in every crypto category. In every category there can be a dozen or more projects to choose from. Let's face it: chances are slim that you are going to pick all the top performers. But if you research the fundamentals, look at sentiment analysis, and read the charts carefully, chances are you can at least pick a handful of winners. All it takes is one or two solid picks to come out in profits.

When you are looking at projects in the same category, and aren't sure which to choose, first compare their market caps to one another, and then ask yourself whether the projects provide the same value to their users. As an example, both Filecoin (FIL) and Arweave (AR) are data storage cryptocurrencies that provide a similar service. Whereas Filecoin is currently ranked #20 with a $9 billion market cap, Arweave is ranked #62 with a $2.5 billion market cap. Assuming that Arweave continues with its current development, it has a minimum theoretical growth factor of 3.6x to catch up to Filecoin. Even if Filecoin continues to outperform, Arweave still has more room to grow.

You can also compare network adoption, TVL, and active wallet addresses. Most crypto projects will proudly display images that show the ecosystem they are creating. Remember: the more genuine network connections, the more valuable the project.

Diversification

The categories in this chapter should be more than enough to get you started, but if you are looking for other possibilities, you can find them on CoinMarketCap under the "categories" tab. Be aware that just because a coin is listed on CoinMarketCap does not mean it is a reputable project. It is essential that you DYOR.

In the same way that traditional investors diversify their holdings across multiple uncorrelated assets in different markets, conscious crypto investors should spread their investments across multiple categories. If you ask your white-haired financial advisor why you should diversify, he'll shoot you a smug glance, puff a ring of smoke from his cigar, and say "risk management."

Let's put this in layman's terms. Imagine that your favorite exchange is suddenly hacked, your champion DeFi lending token is hammered by regulation, or that invincible privacy protocol fails to follow through on its promises. Would you want to be stuck with too big a bag of any of their coins?

Another reason to invest in uncorrelated categories is that there are destined to be some that vastly outperform their siblings. The goal for conscious crypto investors is to ride the **crypto supercars**: coins that yield the highest annual gains in the altcoin index. Who can say for sure whether video streaming, payment processing, or oracle solutions are going to have the greatest utility in the years to come? Since only the shady fortune teller on 9th and Broadway has a magic crystal ball, you're better off putting eggs in at least a few different baskets.

Diversification doesn't mean that you should spread your investments evenly across thirty random coins! If you want to keep up to speed with developments, and be able to sell at the right time, it is best to set an upper limit of projects in your portfolio. For some this may be 3-6, others may feel comfortable with 6-12, etc. If you are new to the cryptosphere, and you really aren't sure how much you can reasonably manage, then it is best to start with a small handful of projects.

Even full time crypto traders who hold more than fifteen coins frequently get into trouble during larger corrections, when virtually every coin takes a nosedive. Even with the most careful planning, when that dip comes, it is like trying to juggle while carrying a pot of water on your head.

Chapter 14

Size Matters, for Sanity's Sake

Before you invest a penny into crypto, you need to decide exactly how much you can afford to spare. In other words, how much of your savings and extra spending money after expenses are you willing to lock into crypto? In order to decide how much you can invest, you first need to calculate your **bank reserve**. You want to create a reserve so that if and when the cryptosphere turns bearish without warning, forcing you to hold your bags for an extended period, you won't lose any sleep.

To determine your bank reserve, first make a list of your monthly expenses. This step is often skipped by new investors who are excited by the possible gains to be made. Monthly expenses include food, gas, rent/mortgage, phone, insurance, credit card payments, and any other regular bills that are non-negotiables for you to stay afloat. Add them together to come up with an average monthly figure.

Exhibit A: Savvy Sophie

As an example, let's say Savvy Sophie has a monthly income of $7,500, with $6,000 in monthly expenses. That leaves her with $1,500 each month to spend as she wishes. Additionally, she has $40,000 in savings that isn't needed for any debts. She knows that crypto is going to play a huge role in the future of finance. At the same time, she knows that the market is volatile, and the bear could turn savage and jump out the window at any time.

Since Sophie's income covers her expenses, if she took half of her savings and invested it into crypto, she wouldn't be overexposing herself to the mean old grizzly bear. Assuming that she chooses not to spend the entire sum of $1,500 on massages and restaurants every month, she could reasonably set aside 10-25% of her spending money to add to her portfolio instead.

Sophie decides to err on the side of caution, investing $15,000 in fundamentally sound coins, and adding anywhere from $150 to $300 to her stack during **flash sales**: sharp, fast corrections when prices are discounted 20% or more. Sophie doesn't have any big plans to cash out her investments, and is allergic to even the thought of reading charts.

Sophie makes a firm commitment to hold onto her crypto for at least 5 years. Because of this, extreme emotions such as greed and fear don't hold sway over any of her decisions. She won't be too concerned when dips happen, or too proud when her investments skyrocket. She will sleep like a baby regardless of the market conditions, knowing that she can rest on her reserves if and when the bear starts clawing at everything in sight.

Fast forward a few years. Sophie's corporate friends, who initially mocked her for taking such a big risk on what they called a "speculative bubble," were shocked to hear about her success with crypto. Over time, many of them realized that they were losing money by simply loaning it to banks. They were eager to convert to her investment strategy, and admired her for her commonsense approach that led to slow and steady long-term gains.

Sophie enjoyed a fairytale ending for three reasons. First, she only invested *after* covering all expenses, and still kept enough in the bank as a safe buffer in the event that her investments tanked. Second, she was careful to only add to her holdings during corrections, steering clear of greed created by price surges. Finally, she had the discipline to hold her coins no matter how the market was behaving. To express their admiration for Savvy Sophie's wins, her crypto peeps gave her the honorific title of **YGMI**: **y**ou're **g**onna **m**ake **i**t.

Exhibit B: Gambling Gus

While this all sounds like a cakewalk on paper, having the discipline and discernment to follow through with Sophie's strategy is sadly lacking in the cryptosphere. Gambling Gus, Sophie's neighbor across the street with three broken down trucks in his front yard, is also into crypto. Oddly enough, Gus has *exactly* the same expenses, income, and savings as Sophie (hey, it's possible).

Gus just got word from one of his drinking buddies that WAGG, the latest puppy coin, is "for sure gonna 100x this month." He decides to **ape** into the project using his entire savings, and includes $3,000 from this month's income to boot. He figures **y**ou **o**nly **l**ive **o**nce, (**YOLO**) and he wants to buy his wife Sally that ATV he saw advertised on insta last week.

To Gus' surprise, WAGG makes a 10x move from 0.001 to .01 in less than a week! He is stoked to see his investment go from $43k to $430k. Only now Gus isn't just dreaming about an ATV–he wants to buy that house in the hills Sally has been pestering him about since they were high school sweethearts. He thinks to himself, "If it already went 10x, why not 100x? I'm gonna wait this out. Then I'll really be a baller."

Tragically, the next day Elon Musk tweets that WAGG creates a tail of code that looks too much like a fire hydrant, which he found while studying their Github repository. This causes WAGG to sag all the way down to 0.00033 in less than 24 hours, one third of what it was when Gus initially invested. To save face, Gus panic sold for a net loss of $28,400. Now not only has he lost his hard-earned savings, he is suddenly in debt, forced to add an additional $1,500 to his credit cards to cover those claw marks on the seat of his pants.

This example sounds pretty outrageous. Believe it or not, it plays out across the globe *every single day.* On CT, people would call this a newbie mistake, made by someone who is **NGMI**: **n**ot **g**onna **m**ake **i**t in the cryptosphere.

Gus never bothered to think about **sizing**: how much capital to invest in different cryptocurrencies. His greed led him to make the classic mistake of overexposing himself. He also played hookie on the DYOR part, and went all in on a speculative meme coin with no actual utility. Needless to say, Gus threw in the towel with his tail between his legs, and his drinking buddies had a lifetime supply of fuel to burn him every time anyone even hinted at the topic of crypto.

Immunity to the Pandemic of Overexposure

Now that you have met Sophie and Gus, take some time to seriously consider who you'd like to emulate. If your answer is Gus, then read no further...have fun in the casino! If it is Sophie, then establish your bank reserve first. If your income doesn't already cover your expenses, and/or you don't have a decent chunk of savings in the bank, then you need to honestly decide whether it is in your best interest to invest. Your sanity and well being should be your top priority in all that you do. Investing money that you don't have leaves you vulnerable to the **pandemic of overexposure**: allocating too much of your resources into risky investments that you can't afford to lose.

I have friends who have refinanced properties, taken out business loans, and maxed out credit cards to buy crypto. They hear about exponential gains from acquaintances and suddenly think they can make a few quick buys, double their money and sell everything to pay off all their debt in one fell swoop. The reality is that it rarely plays out like this, since they don't understand how markets work, and so are unable to properly time their entries and exits. Once excitement goes unchecked, it leads them into a trap. A large majority of the time, they get **wrecked**, (or rekt on CT) losing their initial investment and then some.

Smart money will take the path of least resistance, and steer clear of debt whenever possible. Borrowing from an unknown future is not the way of the conscious crypto investor, who wants to minimize stress while making the most of her/his money. Debts are like parasites–left to their own devices, they multiply at alarming rates. Adding debt adds to your worries. The next time you are tempted to create any form of debt to invest in crypto, first think about how it would feel to have to pay a loan back at up to 25% interest on money that you lost.

If you have reasonably assessed your income and expenses, then it won't be difficult for you to stay out of debt while keeping your bank account balance at a minimum. Smart money is always resourceful. If you don't need to store it in the bank to cover expenses, it should be invested somewhere in the cryptosphere. Investors with an average risk tolerance can put aside anywhere from 6-12 months of reserve funds as a buffer in the event of a bear market.

Chapter 15

CEX vs. DEX: the Good, the Bad, the Ugly

Good Old CEX

A **centralized exchange** (CEX) is a trading platform that allows investors to buy, sell, and swap cryptocurrencies against fiat currencies. In other words, they serve as middlemen who make it possible to swap between fiat and crypto. You pay fees that enable the CEX to cover its operating expenses, to provide liquidity for trading, and to store your crypto.

Advantages

There are several advantages to using a centralized exchange. First, having a central authority behind your capital means that there are actual people to talk to if and when something goes wrong. Customer support can be an invaluable resource for new investors, many of whom feel more secure using a system that resembles what they have grown used to over many decades.

It's also comforting to know that CEXes are insured, so chances are high you will be reimbursed in the event of a hack where funds are stolen. Second, the **Know Your Customer** (KYC) process, during which you provide a government-issued identification, your social security number, and your home address, gives an added layer of security to your account in case you are locked out or otherwise compromised. Third, the trading experience is generally seamless, with options for market and limit orders that will allow you to trade crypto at fair prices.

Disadvantages

Now for the flip side. Centralized exchanges naturally contradict the premise of decentralization. They are backed by for-profit authorities, and therefore must meet similar regulations to modern banks. License and insurance requirements mean less privacy for users. Because they are centralized, they are also vulnerable to hacks, theft, and server downtime.

Assets you store on a CEX are held in **custodial wallets**, which means they have access to your private keys. They can freeze or lock accounts at any time, for any reason. There is also the risk of government shutdown affecting exchanges and all their users, which has happened in China, Russia, South Korea, and Turkey.

Another issue with CEXes is they have full control over which tokens can be made available for trading, deposit, and withdrawal. It can be frustrating to sign up for an exchange, only to find that they don't have a token that you are eager to buy. Or that you don't meet minimum withdrawal requirements. Regularity across exchanges is often limited to the top 10 or 15 blue-chip coins. This means that you will likely need to open accounts on multiple exchanges, meaning more passwords and 2FA tokens to store.

Centralized exchanges can also delist tokens at any time, for any reason. For example, when XRP came under legal scrutiny by the SEC, many exchanges suddenly removed it from their trading panels without notice. Imagine suddenly not being able to buy or sell a token you've owned for several years!

Enter the DEX

Decentralized exchanges (DEXes) are non-custodial trading platforms that enable users to swap between cryptocurrencies without the need for an intermediary. DEXes are built on specific blockchains. Uniswap, for example, is built on Ethereum's blockchain, and allows for swaps between all ERC-20 tokens. Uniswap is considered an **automated market maker**, which is a category of DEX that uses mathematical formulas to set token prices. Smart contracts are used in place of buy and sell orders to link buyers and sellers. The users themselves provide liquidity for others by adding it to **liquidity pools** (LPs) in exchange for a percentage of the fees generated.

DEXes mostly attract investors who want to swap between lesser known tokens that aren't listed on CEXes. All you need to do to use a DEX is launch its app online, connect a compatible wallet with a browser extension such as MetaMask, and choose which cryptos you want to swap. You retain custody of your assets at all times, which means that you are also responsible for your own security. Reputable dexes include thorswap.finance, uniswap.org, sushi.com, and pancakeswap.finance.

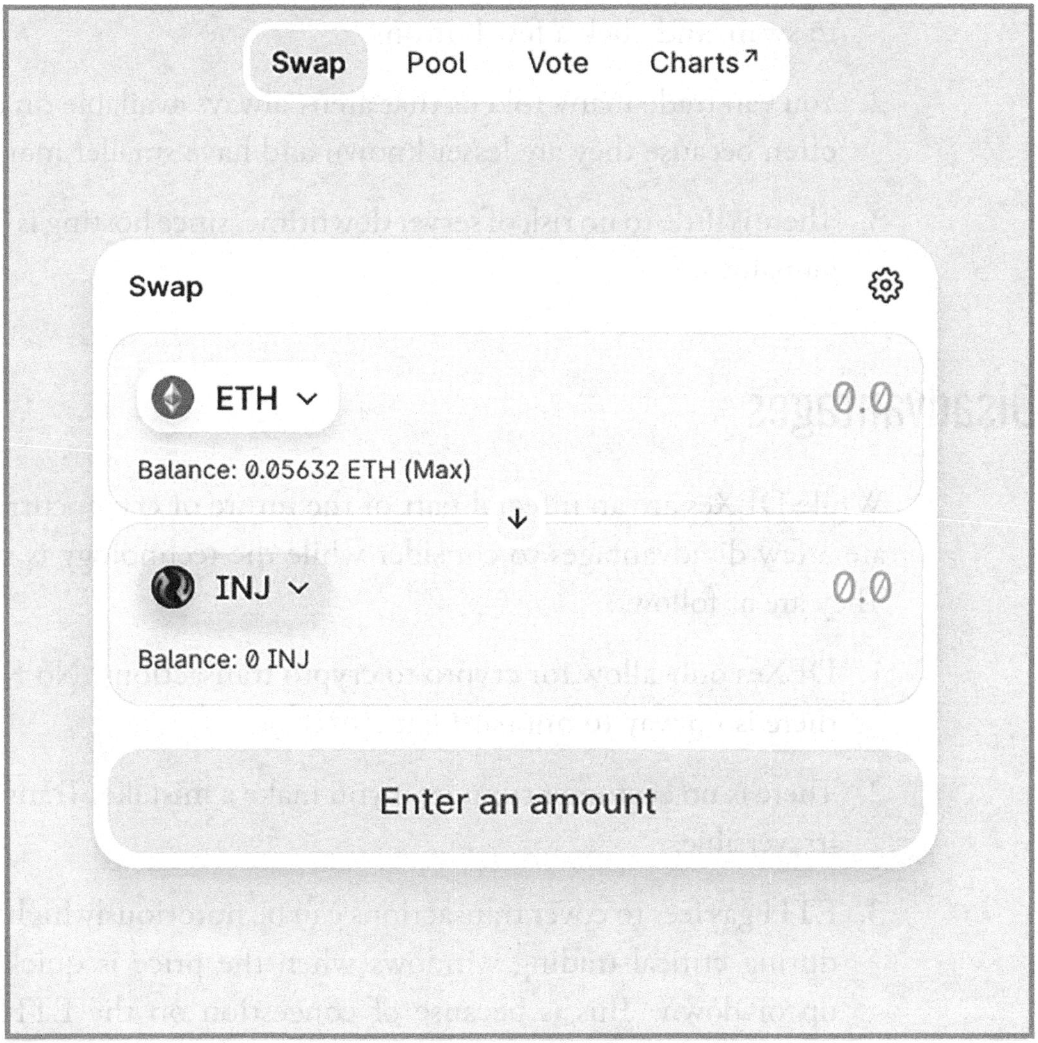

Figure 20: Uniswap's user interface is pretty straightforward.

Advantages

The advantages of DEXes are as follows:

1. There is no need to go through a lengthy registration and KYC process. Simply connect a software wallet, choose the coins you want to swap, and click a few buttons.
2. You can trade many tokens that aren't always available on exchanges, often because they are lesser known and have smaller market caps.
3. There is little to no risk of server downtime, since hosting is distributed globally.

Disadvantages

While DEXes are an integral part of the future of cryptocurrency, there are a few disadvantages to consider while the technology is still young. They are as follows:

1. DEXes only allow for crypto to crypto transactions. No KYC means there is no way to onboard fiat currencies.
2. There is no customer support if you make a mistake. Transactions are irreversible.
3. ETH gas fees to cover transactions can be notoriously high, especially during critical trading windows when the price is quickly moving up or down. This is because of congestion on the ETH network, which also has the effect of slowing down transaction processing. On Uniswap, for example, gas fees range from $5 to $150 or more.

There are also **slippage fees** if there is a difference in the price of two cryptocurrencies between the time a request is made and the swap is completed.

4. Limit orders aren't yet possible.
5. The largest DEXes such as Uniswap and Sushiswap are limited to ERC-20 tokens.
6. DEXes aren't currently as popular as CEXes because they aren't yet as user-friendly, and don't have as high liquidity.

There are DEXes and Dapps that are making it possible to swap tokens across many different blockchains. Automated market makers such as Thorchain (RUNE) will allow users to swap between assets on many different blockchains while paying a fraction of the cost in fees compared to DEXes built on Ethereum. This is a significant development, when you consider that other DEXes currently only allow you to swap ERC-20 tokens.

A Necessary Evil

Despite the potential of DEXes and the inherent flaws of centralization, the CEX model is still a necessity. Over 90% of transactions currently occur on a CEX. This makes sense, since most investors appreciate their resemblance to traditional brokerages. While more seasoned cryptonauts may opt to use DEXes for trading, most people will need time to adjust to the new tech.

Centralized exchanges create bridges between the legacy financial system and the cryptosphere. Users must have the ability to safely and securely onboard their fiat to be converted into cryptocurrency. Additionally, they must have a way to withdraw their funds once they've traded back into fiat. For these reasons, you will need to set up at least one exchange account.

Chapter 16

Getting Started on Exchanges

Before you sign up for the exchange your uncle's friend told you was a safe bet, visit coingecko.com/en/exchanges and take a look at the top 10 exchanges by trading volume. Exchanges with higher volume generally have more trading pairs available, enabling you to swap between cryptocurrencies with more ease.

Remember that exchanges can differ widely when it comes to available trading pairs. You can save time and energy by searching for coins you want to trade before you open the account. Click on the individual exchange links on CoinGecko to see updated lists of trading pairs. Your aim is to minimize the number of exchange accounts you manage for the sake of simplicity.

The next step is to make sure that you live in a country and/or state that will allow you to use the exchange. For example, Binance may see tens of billions of dollars in trading volume daily, but it's not available to US citizens. Binance.us is a separate exchange with only a fraction of the volume and only ⅙ the number of coins. Even then, it's not currently available to residents of Connecticut, Hawaii, New York, Texas, Vermont, Idaho, and Louisiana! Regulations can and will change, but for now these states are throwing crypto temper tantrums. Be sure to search for individual state restrictions as your last check to save time before registering.

Finally, go to cryptofeesaver.com to compare exchange fees. These include deposit and withdrawal fees for both crypto and fiat, and trading fees. On top notch exchanges, trading fees often range from 0.1 - 0.25%, and may depend on your individual trading volume for the month. They are important, but also should be considered alongside trading volume, since higher trading volume indicates higher liquidity and therefore narrower spreads (aka a better deal).

Step-by-Step Guide

The steps below are to help you create an exchange account, secure it using 2FA, and add funds in the form of either fiat or crypto. The process is the same regardless of the exchange(s) you decide to use in your country/region.

Step 1: Sign up using your email, and create and store a unique password with a password manager such as LastPass (covered in Part 4).

Step 2: If you need to onboard fiat currency, you will need to pass KYC checks first. Fill out the required info., and then take photos of the front and back of your license or passport, and upload them.

Step 3: Download Authy app for your phone, and also download the desktop extension. Create an account with an easy to remember master password. Next, navigate to Settings -> Security in your exchange account. Select 2FA, and turn it on for sign in. Open Authy and scan the QR code they give you. You will be able to save the token on your Authy app. Copy and paste the six numbers of the token to verify authentication. Once confirmed, you will be prompted to give the Authy token for sign in to the exchange each time you attempt to login.

Step 4: Now that your account has a high level of security, you can add funds. Exchanges such as Kraken have a "Funding" section where you can deposit and withdraw both fiat and cryptocurrencies. Others such as Binance have a "Wallet" section with the same features. The important thing to remember is that you will need to find either the cryptocurrency or fiat currency that you will be depositing on the exchange, and click deposit.

If you choose to deposit a particular fiat currency, you will likely have several options, and can choose according to your preference. Many exchanges will allow you to either purchase crypto with credit cards, or transfer funds through ACH or by wire transfer. Each has its pros and cons. Credit cards are fast and easy, but the fees can be high. ACH can take minutes to several hours depending on when they are initiated, and you will need to provide your checking/routing numbers on your bank account. They are helpful for transferring considerable amounts without the associated fees. Binance.us, for instance, allows you to send up to $1,000 by ACH per day.

Wire transfer allows you to deposit amounts in the tens or hundreds of thousands weekly. At the same time, it is the slowest and most painful, since you have to fill out extra paperwork. While some banks such as Chase will allow you to do this online, many don't, and there are many numbers and addresses you should double check for accuracy. You may also need to add a very specific reference number in the "Reference" section of the wire form.

The simplest way to get a wire right is to take clearly visible photographs of the information and bring it to your bank. Then when the teller asks you to confirm what you have written on the screen, do it! While double checking the information for the first wire I sent to an exchange, I noticed that the teller had mixed up several letters in the reference section. Note that most banks have a 9 or 10 am cutoff time for wires to go through on the same day.

Chapter 17

Placing Buy and Sell Orders

Cryptocurrencies are destined to be in a perpetual state of **price discovery**, with the price of coins on exchanges in flux as they are bought and sold. The collective perception of the value of coins changes as new information enters the cryptosphere. But there are also other factors that influence the price that you should be aware of before you decide how to purchase crypto.

Most exchanges are fully transparent with their **order books**, which can be viewed within the user trading panel. You will notice that there is a green section and a red section, with the current price landing somewhere in the middle. The numbers are constantly changing as old orders are filled, and new orders take their place. The green section is showing the **bid price**: the price at which different buyers are willing to purchase coins. The red section is showing the **ask price**: the price at which sellers are willing to sell their coins. The difference between the bid and ask price is referred to as the **spread**. Spreads depend on liquidity.

Limit vs. Market Orders

Buyers and sellers use **limit orders**, or offers to buy and sell coins at specific prices, to create liquidity. **Liquidity** is a measure of the ease of buying and selling an asset. The more limit orders there are on the books, the higher the liquidity, and the quicker orders can be filled.

Anyone who places a limit order on the books is considered to be a **market maker**, since they are adding liquidity and therefore helping to stabilize the price of an asset. **Market orders**, on the other hand, are instant buy orders that are executed at the current market price, removing liquidity. A **market taker** is someone who uses a market order to take away liquidity by either buying or selling. Day traders who need to make immediate trades would be one example. Market makers pay lower fees than market takers, since they serve to narrow the spreads by adding liquidity to the order books. Market takers pay a premium for the convenience of instantly having their orders filled.

Understanding liquidity is important, since different exchanges have different spreads based on available liquidity. Lesser known exchanges will have higher spreads, which means you risk paying more than you would on other trading platforms.

Another factor to consider before placing orders is the illiquidity of lesser known coins across exchanges. Less supply and demand means there will be less orders on the books. This creates slippage that can be especially problematic with larger market orders. A 1% slippage on a $10,000 market order means the total cost for the asset will be $100 different than expected.

As another example, say you have illiquid coins that are suddenly plummeting. You may plan to sell them, but there is such a discrepancy in the spread that your orders aren't filled, leaving you stuck with your bags. Whether buying or selling, the less liquidity for a coin on an exchange, the more unstable the price. When in doubt, navigate to the specific trading pair for the exchange you use on CoinGecko and take a look at its spread.

#	Exchange	Pair	Price	Spread
*	eToroX Sponsored			
*	Crypto.com Sponsored			
1	Gate.io	AR/USDT	$59.17	0.05%
2	Binance	AR/USDT	$59.28	0.15%
3	HitBTC	AR/USDT	$58.98	0.36%
4	Huobi Global	AR/USDT	$59.20	0.85%
5	MEXC Global	AR/USDT	$59.33	0.19%
6	Binance	AR/BTC	$59.23	0.02%
7	MEXC Global	AR/ETH	$58.94	0.3%
8	BitMart	AR/USDT	$59.39	0.57%
9	Binance	AR/BUSD	$58.27	0.33%
10	CoinBene	AR/USDT	$58.93	0.78%
11	Hoo.com	AR/USDT	$58.17	0.64%
12	CoinEx	AR/USDT	$58.06	0.09%
13	Huobi Global	AR/BTC	$59.07	3.85%
14	Binance	AR/BNB	$58.22	0.68%
15	Bittrex	AR/BTC	$59.03	0.74%
16	BigONE	AR/USDT	$58.12	0.84%
17	CoinEx	AR/BTC	$57.67	7.69%
18	BHEX	AR/USDT	$57.67	1.25%

Figure 21: Notice the difference in both the price and spread for AR/USDT across different exchanges. CoinEx has a 7.69% spread!

Limit buy orders are bids that you place below the current market price, hoping the price of the coin will drop and your order will be filled. For example, if I want to purchase $100 worth of RUNE for $11.11, but the current price is $12.36, I can place a limit order with $11.11 as the price I'm willing to pay. Limit sell orders work on the same principle, except the ask price you are willing to sell at will be higher than when you purchased. For example, after my limit buy order filled at $11.11, I could set a limit sell order which would fill when RUNE hits my ask price of $27.36.

The risk with limit orders is that the price may continue to run up or down, leaving your bid unfilled. The reward is that if the price drops to your bid, you will receive a discounted price. Likewise if it rises to your ask, you will sell at a profit.

All limit orders will show up in your trade history as open orders. The money you have in your account will be docked for limit buy orders until the order is executed. In other words, you won't be able to use those funds until the order is either filled, or you cancel it manually. To cancel an order, it's usually a matter of clicking a checkbox on the far right in the "open orders" panel.

Note that certain crypto apps such as crypto.com and cash app don't give you the option to place limit orders. Crypto wallets such as Trust wallet and Ledger also offer quick buy options, but with the same caveat. You also may not be able to use ACH or wire transfer to onboard your fiat. With credit card fees as high as 10%, this is something to consider before registering. In all of these cases, lack of functionality is the sacrifice you make for convenience and simplicity.

Instant buy and sell trading apps are basically crypto convenience stores. They'll get you your quick fix without all the bells and whistles, but you're likely going to pay more due to larger spreads and higher fees. If you are buying larger amounts over time, you will save a lot more money using a vetted exchange. Exchanges such as FTX, Kucoin, Kraken, and Binance have great apps with full functionality and lower fees.

Take Profits and Stop Losses

3Commas is a command center for all types of trades. It links to many top exchanges through APIs, and can execute trades on your behalf without you even having to login to your exchange(s). It will also tell you exactly how much money you are up or down on each trade, which is sadly absent on exchanges. There is even a mobile app for those who need to track profits and losses and make trades on the go.

The best feature of 3Commas is smart trading. You can set limit orders, take profits, stop losses, and more. **Take profit** is an order for a trading platform to automatically close part or all of a trade on your behalf at a predetermined price.

Stop losses are similar to take profits, only they are used to prevent losses rather than lock in gains. When you set a stop loss, the trading platform will automatically execute a sell order for part or all of your position once the price drops to the preset level.

While you can also set stop losses and take profits through traditional exchanges, 3Commas will allow you to create different tiers where a specific percentage of the coin is bought or sold (Figure 22). As an example, let's say you bought RUNE for $7. You could then set the following take profit levels: sell 20% at $10, sell 45% at $15, and sell the rest at $20.

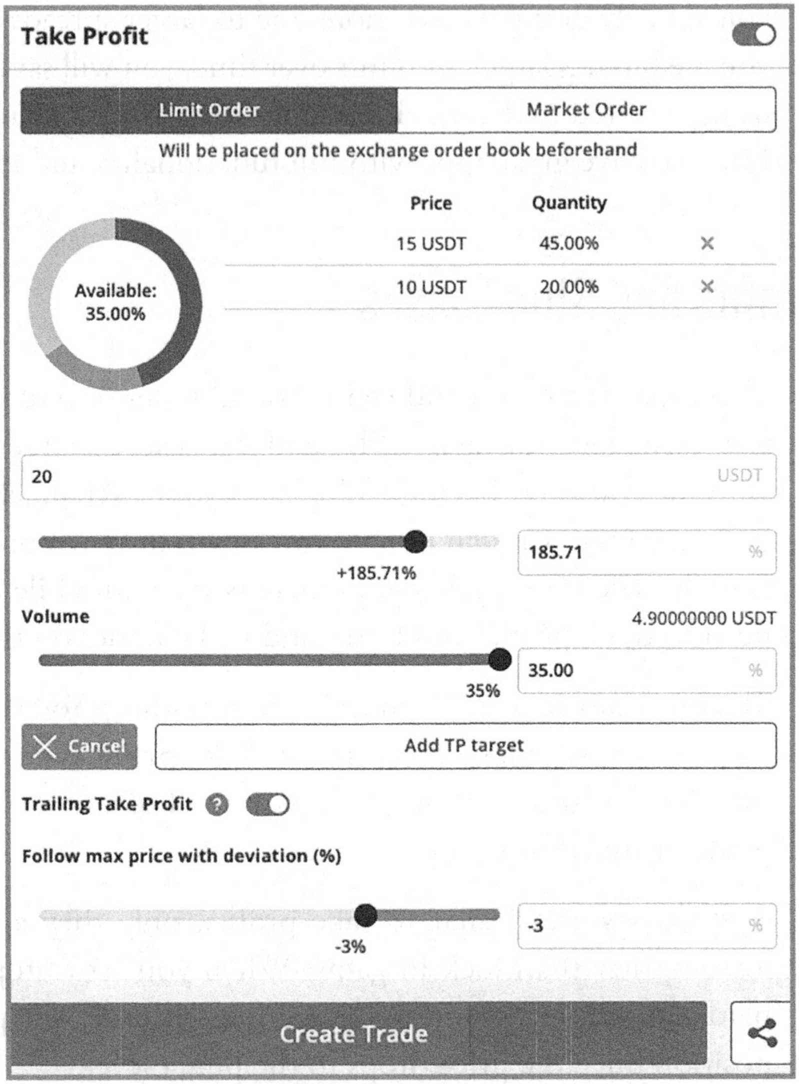

Figure 22: Setting profit targets for RUNE using 3Commas. Notice that "Trailing Take Profit" is activated for the last target at $20, with a 3% follow max price.

Another great feature is the option to set trailing take profits and stop losses. Let's say that for your last take profit tier in the example for RUNE above, you set a trailing take profit of 3% at $20. Once the $20 strike price is hit, the trailing take profit is activated. If the price suddenly drops 3%, the sell order is executed. But if RUNE continues to rise, your coins won't be sold until the price drops 3%. Theoretically, this means that RUNE could skyrocket to $33.3, and you could still be in the trade, given that there were no 3% pullbacks along the way.

Trailing stop losses are similar (Figure 23). If you bought RUNE at $7, after setting all of your take profit targets, you could then activate a trailing stop loss at -10%. This would sell your position if the price dropped to $6.30. Then as the price rises, the stop loss will rise as well, so that it always remains 10% below the current price. If RUNE rises to $10, the trailing stop loss will close your position if the price drops to $9.

Figure 23: The stop loss feature on 3Commas. Here a trailing stop loss of -10% is used.

Paper Trading and AI Trading Bots

3Commas also offers a paper trading account with a balance spread out over various cryptocurrencies so that you can practice trading with paper money. The option to give your trading strategy a dry run can save you boatloads of money when it comes time for the real deal. You can use all the standard features of 3Commas, the only difference being you can't cash out when all is said and done!

Professional traders use paper trading accounts to test out AI trading bots that they have developed. **Trading bots** are computer programs that execute trades for you based on certain parameters that you set. As opposed to smart trading, trading bots will continue to buy and sell different cryptocurrencies as long as they remain active. As an example, 3Commas has grid bots that are used when the price is consolidating in a specific range. The grid bot will create buy and sell orders as the price bounces up and down in that range.

AI trading bots are increasing in popularity since they remove emotion from the trade. At the same time, they require constant monitoring, and aren't practical for long-term investors. Bots execute hundreds, even thousands of much smaller trades, which complicates taxes and creates higher trading fees on exchanges. Everyone is different–if you want to see how they work, try them out using a paper account on 3Commas.

Sticky Trading Pairs

You will notice that most coins have multiple different pairs. A **trading pair** consists of two assets that are traded together. BTC/USD is the simplest example, since it shows the price of bitcoin in terms of dollars. Almost every major coin can be paired with BTC and ETH. It can be confusing to use certain trading pairs, because the value is relative to the two assets in the pair. For example, if you search for XMR/BTC, you will see the price of 1 XMR in terms of BTC. Since 1 BTC costs exponentially more than 1 XMR, the price becomes a small fraction. As of this writing, 1 XMR costs about 0.00734 BTC.

Trading pairs are useful in situations where you want to rotate gains directly from one coin into another. For example, if BTC recently rose 30%, and you see that ATOM is showing signs of going stratospheric, you can use the ATOM/BTC pair to sell your BTC directly into ATOM, saving you the fees from having to convert to USD first.

For most other trading scenarios, it's easier to start with either USD or a stablecoin. **Stablecoins** are cryptocurrencies with a value that is pegged to a fiat currency such as USD, another cryptocurrency, or a commodity such as gold. They are backed by reserve assets, and are able to offer general price stability in an otherwise volatile market. Tether (USDT) is a stablecoin that is pegged to the dollar. 1 USDT = 1 USD, with minor variations depending on supply and demand.

Trading in pairs outside of USD and stablecoins gets complicated when it comes to limit orders. If both coins are volatile, it is tricky to pin down a specific target for the coin you want to buy. It can be frustrating to not have a limit order filled because of an unexpected discrepancy in the price of the two coins.

Another issue with unconventional trading pairs is that it complicates taxes. Regardless of your chosen trading pair, all sell orders made less than a year after purchasing crypto fall under the general umbrella of short term capital gains. This means that you will need to translate the numbers into dollar amounts at the time of trading so that you can determine how much you earned, and how much you are reinvesting.

To keep things simple, consider sticking to the stables as much as possible. If you need to do a quick swap, it's no big deal, but know the risks and pitfalls before you trade.

Chapter 18

Trading with Debt?

The Siren Named Leverage

Fair profit comes from fair dealing. Remember this mantra as we explore the concept of leverage, which is a popular trend in the cryptosphere. **Leverage trading**, also referred to as margin trading, is using borrowed funds in an attempt to amplify gains in a short time period.

For example, let's say our friend Silas would like to invest $100 in a leverage trade. $100 is his **initial margin**: the collateral he is willing to put down to receive a loan from a participating exchange. Many exchanges offer 3x, 5x, 10x, all the way up to 125x leverage. Silas decides to use 10x leverage, so he receives $1,000. In return, he will pay interest to the platform. Binance, for instance, charges a daily management fee of 0.01%. If he takes a **long position**, then he is betting that the coin will rise in price, while a **short position** means Silas is betting the coin will drop in price.

Sounds like a sweet deal, right? Well, it depends which way the coin flips. If Silas bets long, and the price of the coin goes up 1%, he will gain 10%. But if he is wrong, and the price drops 1%, he loses 10%. More importantly, if the price of the coin drops 10%, the exchange will automatically **liquidate**, or close Silas' entire position, keeping his initial margin. If they didn't perform the liquidation, they would stand to lose more than Silas, since there is no longer any collateral to back the loan.

The higher the leverage, the higher the risk. To determine how much of a move it would take to lose the initial margin, you can simply divide it into 100. Using 3x leverage, it would take a 33.3% move in the opposite direction to lose everything. For 125x leverage, a miniscule 0.8% move the wrong way liquidates the position!

In an emerging, and therefore highly volatile market, using leverage is like trading on cocaine. In general, it is highly attractive to addictive personalities and extremists who are obsessed with making a quick buck. Sure, there are some professional traders that have done quite well with leverage trading. Remember that your chances of winning or losing don't change when using leverage. What changes is the *degree* to which you win or lose. This creates unnecessary stress, and a faster game that requires more of your time and energy.

When you trade on leverage, you are essentially borrowing from an unknown future. But the conscious crypto investor's goal is financial freedom, which means the steady elimination of traditional debt. Moving with the flow means focusing on the timing and size of trades to minimize risk. This is a zen mentality that will seem slow and boring to the impatient. But those who are patient are happy to steadily grow their investments with trading strategies that prevent overleveraging.

Credit Cards

It isn't helpful to purchase crypto with credit cards. Even if the fees on the platform are low or free (for a limited time), you are borrowing money from the bank that issued the credit card. You will pay up to 25% interest, which means that your investments will have to increase *at least* that much annually to pay off the loan. Then you must also consider capital gains taxes for short term sales, which typically range from 15-40% depending on your income. Credit cards entice you to add to your debt with reward points and miles that in reality don't come anywhere close to the interest you must pay back. Remember that if the terms for credit cards favored you, then they wouldn't have offered them in the first place.

Chapter 19

Should You Join the HODL cult?

HODL was created by a cryptonaut who was sipping lemonade from his porch while trying to talk his crypto peeps down from panic selling as the bears started their free fall. He tweeted, "Don't sell your bags, hodl them like granny on the subway!" They started to panic even more, so he clarified: **H**OLD **O**N **F**OR **D**EAR **L**IFE! (Actually someone just misspelled hold, but this sounded better)

Arguments for HODL

HODL is arguably the only crypto cult that would receive granny's seal of approval. It also happens to have the most appeal for the average investor, for several reasons:

1. **Investors Have Lives Too**

Most people who are invested in crypto already have full time jobs, families, and a weekly laundry list of responsibilities and social activities. They don't have time to stare at a computer screen for hours every day, tracking the movements of their favorite coins. What's more, the crypto markets are open 7 days a week, 365 days a year. Trying to catch every significant rise and fall is impossible, even with alerts from price tracking apps.

Let's face it, most of us already spend way too much time on our phones. Do you really want to be putting in overtime trying to stay up to speed with all the latest developments of a technology that is growing faster than the internet?

2. Skip Out on the Emotional Rollercoaster

The odds are stacked against traders. Even full time professional traders who have been dealing in crypto for over a decade lose large sums of money buying and selling. CT is their support group!

Remember those rides at the carnival that spin you in circles until you can't see straight? Crypto trading begins to feel like that after a while. If you want to trade consistently, it demands a great deal of your time and energy on a daily basis. That includes the weekends, if there is a sudden spike or drop in the price. You have your up days, and your down days, and you mostly get off the ride feeling pretty woozy.

Some of the most successful crypto traders I know don't have other jobs, kids.....even pets! They are usually drinking some form of caffeine, and it can be hard to tell the last time they got a good night's sleep. Is this the type of lifestyle you want to commit to?

3. Can Ignore the Charts for Years at a Time

When you invest in a coin with strong fundamentals, it's easy to HODL. You have faith that the price will go up over time. If you believe that the price of bitcoin will be 1 million dollars in five years time, then why all the fuss about the price dropping from $60,000 to $30,000? HODLers don't need to learn sophisticated TA, disrupt their schedules, or worry about how much the price has increased or decreased. They simply **set it and forget it**.

4. Avoid Capital Gains Tax

It is well known that if you don't sell your crypto, you don't have to pay any taxes. Short term capital gains on crypto sales are taxed at the same rate as ordinary income, which in some cases is over 40%. After just one year of HODLing, you will pay significantly less–between 15 and 20%–for the same gains. This also makes it possible for HODLers to loan out their crypto, and only pay taxes on the interest gained each year.

Swing Trading: A HODL Alternative

While HODLing is statistically the most successful trading strategy for the average investor, by reading this book you are on your way to becoming a conscious crypto investor, and should at the very least weigh all possible options. The set it and forget it HODL strategy assumes that investors have A. no interest in taking risks with short term trading, and B. little to no knowledge of chart patterns or market psychology. Since they will likely be unable to keep their emotional impulses at bay, the less involved, the better.

While HODLing spares investors from trading mistakes, and keeps them free and clear of short term capital gains taxes, there are also several major disadvantages. The first is the huge drawdowns that take place after every bull cycle. After the 2017 bull market top, the price dropped 84% from it's **all time high** (ATH) before slowly making its climb back up again. It took three whole years for the price to return to its previous ATH.

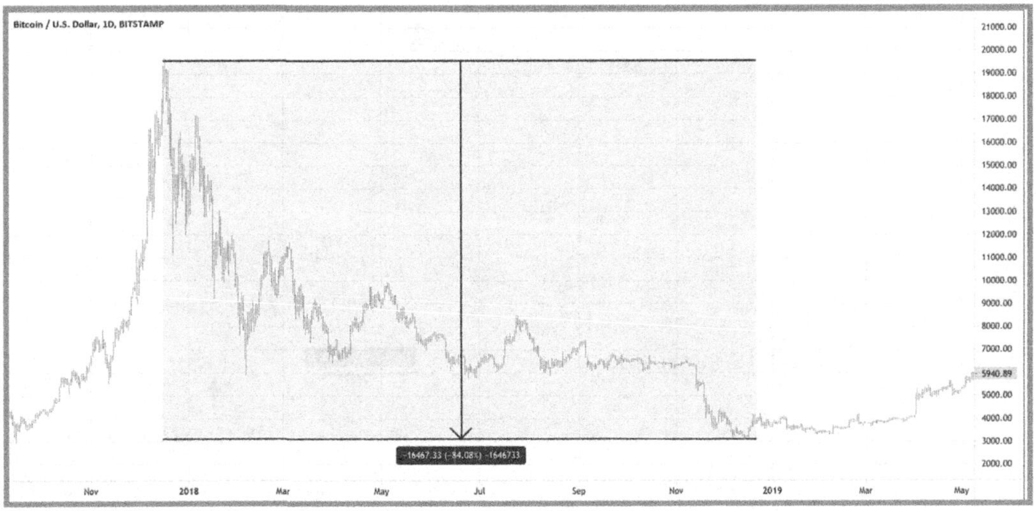

Figure 24: Bitcoin dropped 84% during the 2018-19 bear market.

Imagine for a moment that you bought bitcoin in December of 2017, when many new retailers rushed into the cryptosphere while the price for 1 BTC was between $15,000 and $20,000. Would you have invested at that time if you knew that you'd have to wait several years to even begin to see any return on your investment? Probably not. Would you have even wanted to HODL after a 50% correction? Not likely.

Now let's assume that you manage to buy in shortly after the 2014-15 bear market ends. You would have conservatively bought bitcoin for about $330, when the price began to steadily rise after shrugging off a long downtrend. In this case, even if you completely missed the opportunity to sell all the way up to $19,500 in 2017, and you made the mistake of selling the absolute bottom of the next bear market at around $3,100, *you would still be up 9.5x.* Can you see why timing your entry and exit is so important?

Figure 25: Buying at the right time can make up for some serious mistakes, like selling at the bottom of the next bear market. Worst case scenario: 9.5x gain in just over 3 years.

HODLing can be advantageous given that you understand price trends (covered in Part 3). But most investors don't understand the first thing about when to enter a trade. Those who are able to buy low could theoretically apply the same methods to sell high. If you had the option to buy low and sell high every time, would you take it?

Let's take the last example a few steps further. After investing $1,000 into bitcoin in August of 2015, you saw the steep parabolic rise in

December of 2017 as a warning sign of a bull cycle top. You therefore decided to sell some or all of your bitcoin at around $16,000, which is $3,600 less than the cycle top of $19,600. This would leave you with over $48,000 in realized profit. You held your investment for over two years, and therefore paid long term capital gains taxes of 15%. Your net gain is $40,800, an almost 41x **return on investment** (ROI).

Figure 26: A classic example of buying low and selling high. Note that $330 was still 116% higher than the bear market bottom, while $16k was 18.5% lower than the bull market top.

Next, you wait out the 2018-19 bear market until the downtrend starts to show clear signs of a reversal. You buy back at around $4,800 per BTC on April 3, 2019 after a strong daily close on high volume that breaks out above a two year bearish trendline. You now own 8.5 BTC, which is 5.5 more BTC than if you had simply HODLed. Was selling the top and paying the $7,200 in taxes worth an extra 5.5 BTC? Let's revisit that question in 10 years.

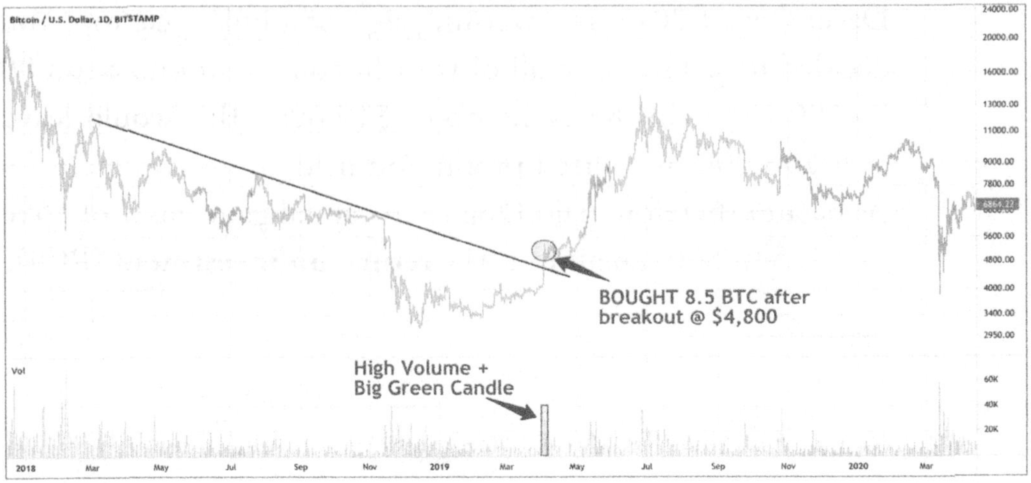

Figure 27: Even waiting to buy back after a confirmed breakout would have yielded you 8.5 BTC, assuming that you invested all of your previous earnings after taxes.

Most HODLers will argue that the likelihood of spotting exact tops and bottoms is slim to none. They are 100% correct. But the buy and sell prices you see in the example above aren't tops and bottoms–they are at clearly defined breakout and breakdown price ranges both before the top, and after the bottom.

This may all sound like a foreign language to you right now. Part 3 will break down these concepts in much more detail. For now, the takeaway is that even though the gains were exponential, they were actually based on conservative **entries** (buys) and **exits** (sells) after several clear signals. If you had made the above moves, you could have earned 5.5 BTC in just three trades that spanned almost 4 years!

Chapter 20

Other Trading Strategies

The DCA Debate

Dollar cost averaging (DCA) is the practice of gradually investing in one or more assets over weeks, months, or years. This is done to reduce risk, since price fluctuations are common in all markets. It also has the effect of eliminating emotion from the equation, since it is a long term strategy that isn't as concerned with how the market is behaving on a day-to-day basis.

While DCA is attractive to the average investor because of its simplicity, it is not without its limitations. First, while your risk is spread out over a longer period of time, it isn't always reduced. For example, say you invest $100 in BTC on the 6th of every month. What if the price of BTC happens to surge on that day for four consecutive months, only to drop back down again after? Of course it is also possible that the price falls each month on the 6th, and rises again the next day. Either way, this strategy randomizes your chances of investing at the most opportunistic times. There is no consideration of market sentiment, fundamental analysis, or technical analysis. You might as well be throwing darts at the charts!

This may also cause you to miss out on higher returns over time. If you have $10,000 to invest now, why spread it out over 8 years? Assuming the project is fundamentally sound, even in 5 years the price will likely be exponentially more than it is today. While you obviously don't want to ape in when the price is **mooning**, (going up fast) the odds are in your favor that the price you paid today will be a bargain five years from now.

DCA can be beneficial in situations when it is timed properly. If you have $100 extra after expenses every month, and you know how to read charts (you will soon), you can plan your entries to coincide with market corrections. For example, if the price is skyrocketing for three months straight, you could wait patiently until a 40% drop in the fourth month to invest all $400 at once. This strategy is called **buying into weakness**.

Another DCA strategy that is popular with many traders is called **doubling down**: buying twice as much with each successive entry as the price dips. If you spend some of your reserves to purchase a coin for $10 after a 20% correction, you would spend twice as much to buy it at $5, and then again double down at $2.50, and so on until you either run out of reserves or the price begins to increase again.

Even though DCA is technically used as a strategy for buying an asset, crypto investors sometimes use DCA to refer to **scaling out**: gradually selling a position. Selling part or all of your position as the price of an asset is increasing is called **selling into strength**. Although it takes patience and discipline to DCA properly, gradually buying into weakness and selling into strength will lead to much greater returns on any timeline.

While DCA is appealing to many because of its seeming simplicity, in truth it is an art that must be mastered with practice and a careful consideration of market conditions. When done randomly, DCA becomes a crutch for the average investor that keeps them from learning the fundamentals of trading. Conscious crypto investors *always* make informed decisions based on careful observation.

Day Trading

Day trading deserves mention, only because it seems to have an especially strong appeal to those who are new to crypto. What seems simple at first glance is actually more complex than many realize. To put it simply, day trading is like investing at 3x speed, which means less room for error. A large majority of successful day traders–those who can consistently turn a profit over time–have years or even decades of experience with trading, and are highly proficient at pinpointing price patterns. For everyone else, it's like playing the slot machines: the game is rigged, and while you may make a few winning trades, sooner or later the scalpers will come for your money.

The main advantages are that you can A. enter and exit trades relatively quickly, and B. have the peace of mind knowing that you will be in cash more frequently. If you can stomach the emotional rollercoaster of watching the price rise and fall, and you don't mind staring at charts on your computer for several hours a day, then it is something to consider.

But day trading is a high adrenaline activity that can be *very* stressful, since you risk losing much more money in a much shorter time period. It requires more precision, and since you are working with shorter timeframes, chart patterns can be tricky to confirm with accuracy. Even if you are limiting day trades to specific hours and days of the week, it requires your full attention, and can be mentally exhausting.

Chapter 21

A Conscious Investor's Rules for Trading

1. Pocket Some Dry Powder

Consider keeping a stable fund of at least 10% of your total investment capital at all times. An even safer approach would be to stable 15-30%. This is the amount of crypto you have converted to a stablecoin such as USDC, UST, etc. Holding stablecoins is often referred to as "having dry powder," since they are a form of cash reserves. When every other coin under the sun is sinking like a stone, you will be glad your stablecoins aren't being dragged down with them!

Even if you have plenty of money in the bank, and are willing to risk losing your entire crypto investment, you should still keep a stable fund. Aside from the protection this affords you during inevitable and sometimes frequent market corrections, an even greater advantage is the option to use a limit order strategy to benefit from fire sales when prices drop. Sometimes during a flash crash, you will see a very long lower price wick, where the price dips momentarily but the supply is instantly bought (Figure 28). When this happens, you can be certain that these were limit orders set by smart money, who anticipated the dip and planned accordingly.

Figure 28: A longer wick during a correction where buyers immediately absorbed the supply from sellers, likely using limit orders.

If you are 100% invested all the time, thinking the price can only go up, you have made the mistake of 90% of investors. Sheep buy impulsively, missing out on large wicks where prices can sometimes drop 20% or more in less than an hour! If you follow the limit order buying strategy, There is no greater feeling than having a low bid filled, followed by an immediate reversal in price. This is where the greatest gains can be made.

2. Never Buy an All Time High

Buying a coin that has just made an ATH can be a risky move. Price has memory, and charts are visual displays of emotion over time. When trends are broken and history is in the making for a coin, anything can happen! It is true that the price could continue to pump, with some coins doubling or even quadrupling after reaching an ATH. But if you zoom in close enough to charts that show these parabolic rises, you will see at least one drawdown back to a previous level of support.

Imagine a bird that has just learned to fly. After leaving the nest, even though it knows its own abilities, it may return to its familiar home several times. This happens because the bird must mentally digest the experience. Eventually it will make its final departure, never to return to the nest. But in the beginning, it clings to the comfort of what it has known.

Your task is to be the shadow of a coin (bird), closely tracking its rise and fall. Remember that cryptocurrencies are entities that embody the aspirations of their participants. The ups and downs in price they experience are both healthy and necessary for their growth. **Parabolic** rises–steep ascents into uncharted territory–must be balanced for the overall well being of the coin. This is no different from an energetic dog that must rest after hours of play at the park.

As an investor, it is your job to not only anticipate periods of reprieve in the markets, but to use them to your advantage. Buying into a dip during a longer term uptrend is a smart money move that every conscious crypto investor makes.

Even though momentum is in a coin's favor during an uptrend, as a buyer you want to buy in the **hiccup** of a chart: a pause where the price consolidates before making its next big leap. If you are patient, you can take your position at a level of previous support, and then ride the wave up.

3. Never Buy or Sell Your Entire Bag at Once

The conscious crypto investor is always patient, and will not act impulsively. Buying and selling all at once is an impulsive move, one that you will regret 9 times out of 10. The reason is simple: volatility. Remember that the price of any given coin can fluctuate wildly at any time. When trades are made spur of the moment, you leave yourself no wiggle room when the price makes a significant jump either to the upside or the downside.

In ancient Daoist philosophy, the emphasis is on finding **the middle way**, not leaning too far to either extreme in anything that we do. The trick is to find your **center point**: the place of grounded awareness that helps you to make smart investment decisions. If you can stay equanimous amidst all the violent swings life throws on your path, then you will be better equipped to weather the storms. You will also have a greater appreciation for the calm when all is going smoothly.

Mastering the art of scaling into and out of your positions should be your top priority as a conscious crypto investor. When there are radical swings to the upside, one who follows the way will lean back into the middle. This means selling some of a position, and then waiting patiently for the price to return to a reasonable level of support. If you follow this practice, you will be well poised to buy back in at a lower price. Your patience and adherence to the middle way will reward you.

When there are sharp swings to the downside, a follower of the way will again lean back into the middle. In this case, this can mean one of two things: 1. having a preset stop loss that sells some or all of your position when the price drops below critical support levels, or 2. waiting out the storm. In a bull market, the simplest, most effective strategy is to wait patiently for the uptrend to continue. This assumes you are invested in a reputable project that has a track record of strong performance.

4. Consider a HODL Fund

Consider allocating 10-30% of your investment to a **HODL fund**. Stick to it regardless of market conditions. For example, if you invested $1,000 in a project, you would commit to HODL anywhere from $100 to $300 for at least 1 year. You can of course HODL for as long as you'd like, but making a firm commitment is what is important.

Even though we already saw the greater potential reward of well-timed trading, the truth is that even in the most bearish market, you really never know how low the price will drop before reversing. By retaining some of the asset no matter what, you mitigate the risk involved with going extreme and selling everything. Think of it as your safety net: a portion of your crypto that you have predetermined will not be influenced by your emotions.

If you have followed the formula for determining meaningful projects, and you truly believe in what they are creating, then you will have no problem keeping at least a small percentage of their coins active in your portfolio for a year or more.

5. Take Profits While the Market Is Hot

Taking profits, often referred to as "skimming off the top," means realizing your gains by selling into strength. This can be in the short, medium, or long term. For example, one strategy would be to sell off your initial investment once it doubles, and then set progressively higher take profit levels as the price continues to rise. This minimizes your risk, since what is left is house money (pure profits). This also lowers the emotional aspect of the equation, which may lead you to make smarter trades in the future.

Taking profits sounds easy–that is, until you try to put it into practice! If you invest $100 in ETH today at a price of $4,000, and three months from now the price of ETH rises to $8,000, you will have doubled your investment. At this point, the average Joe is thinking, "I just doubled my investment! If I take it out now, I might miss out when it doubles again!" This mentality plays out to the ultimate demise of legions of investors every time a coin pumps.

To give a tangible perspective, take a look at the chart for DOGE in Figure 29. Each of the huge dips are cliffs for sheep. The wolves will lead them to the edge, and then quietly back away and wait patiently. The sheep bought as the price was increasing and didn't sell in time. The wolves, who were on the leading edge, bought much earlier and then sold all the way up to the top!

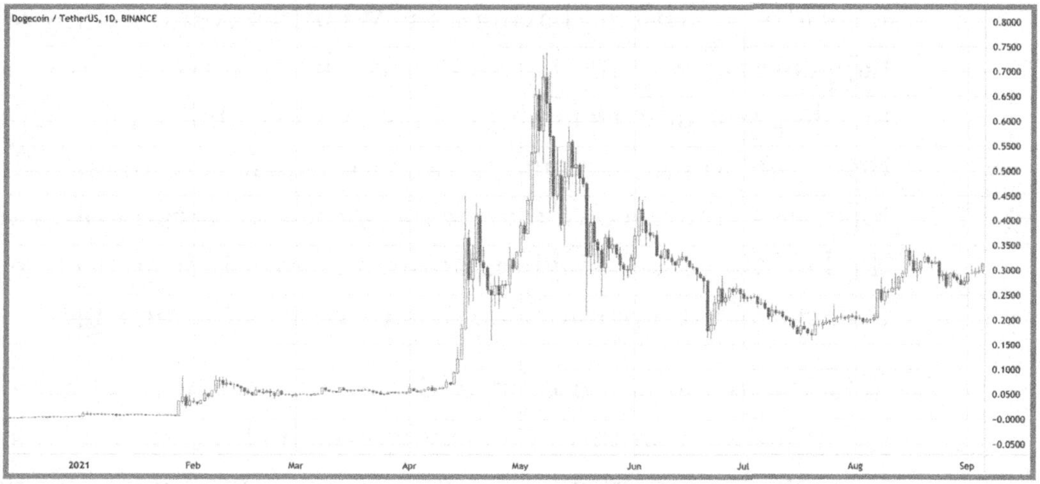

Figure 29: Notice how sharply the price fell after each parabolic rise. Wolves were selling while sheep were buying.

6. Have WOAH, not FOMO

Fear **o**f **m**issing **o**ut, or **FOMO**, is one of the greatest driving forces in all markets. FOMO is like walking into a dark tunnel with your friends because they said there was a secret surfer island on the other side with 99.9% females, personal coconut attendants, and the glossiest waves to shred year round (OK, so not everyone has the same FOMO weaknesses).

FOMO is herd mentality, and the chances of getting burned are much higher than ever profiting. The reason is simple: FOMO only exists when the market is hot. Remember March of 2020? Buying bitcoin was so unsexy, it was the *last thing* on people's minds. And yet, it was the investment of a lifetime. Don't think that time will come again? That's FOMO tunnel vision, my friend.

Instead of FOMO, try having some **WOAH** instead: **w**aiting **on a**wesome **h**appenings. WOAH fanatics buy when everyone else's portfolio is bleeding (steadily decreasing). They may even buy while they are licking their own wounds, doubling down on their investment. The WOAH folks are so outrageous, they *do the opposite of everyone else.* Even though the FOMO gang is laughing at them nervously from deep in their dark tunnel, WOAH investors are secretly on the leading edge.

WOAH investors are calm and patient. The price of bitcoin can be going up for months, and they won't flinch. They will wait patiently for a **black swan event**: a negative event or announcement that catches investors off guard, increasing sell pressure and leading to a systematic correction. During these times the price will unexpectedly drop 50% or more. WOAH folks will watch it bounce, and methodically buy using their reserves during each dip. They will set low bids at key support levels, and wait to see if they fill. While FOMO fanatics are breaking bitcoin clocks and screaming at the charts, WOAH cats are purring in their sleep.

7. Keep a Trade Journal

Even though every trade is recorded on the exchange, actually taking the time to write down the date, amount purchased or sold, and your gains or losses will leave more of an impression in your mind. Keep a separate notebook with well-organized columns that you can easily fill in. Rather than obsessing over gains and losses, include a section with notes about important lessons learned. Be kind to yourself. These notes will be a valuable tool moving forward.

8. Create a Reasonable Crypto Routine

You realistically only need to spend 10-30 minutes daily absorbed in the cryptosphere. As an example, you might spend 5-10 minutes studying charts, 5-10 minutes reading crypto news, and the rest of the time learning about a project. Every day will be different. Balance crypto days with the inevitable holidays: weekends or vacations when you don't pay any attention at all!

This approach assumes you have a clearly defined trading strategy. It is also helpful to set a few important price alerts for coins in your portfolio on a tracking app such as Coinstats to be notified of big price moves. If you have a hard time committing yourself to 10-30 minutes a day 3-5x per week, then your safest bet would be to stick to a long term HODL strategy. Knowing that you won't be buying and selling for a year or more protects you both financially and emotionally when big dips happen without you knowing (they definitely will!).

Maybe you have the opposite problem, spending too much time reading useless banter on Twitter or impulsively listening to speculative crypto analyses on YouTube (Bitcoin to 10 million in blah blah). You may notice that especially during euphoric bull runs, you tend to get wrapped up in the excitement. While a little excitement is a good thing here and there, if you aren't careful, it can turn into an obsession that leads to burnout quicker than you can say Satoshi.

If you are winning in crypto but losing with your family, health, and/or mental stability, then you are losing the game of life. Crypto should be treated as one of many extensions of your vital energy. It should *never* be prioritized above physical, mental, emotional, and spiritual aspects of life. In fact, crypto should be considered last, after you have done your best to balance everything else.

9. Give Yourself a Break

Investing has a lifelong learning curve. If you start out thinking that you will master the art in a few weeks or months, you are going to get tossed around in the waves. Impatience leads many retailers running circles in the revolving doors of the cryptosphere, constantly doubting whether to stay in the game.

A better approach is to maintain a beginner's mind, taking the slow and steady path while appreciating lessons learned along the way. Give yourself permission to make mistakes, and don't be in a hurry to learn everything all at once. If you aren't having fun, change the way you play the game!

Part 3
The Skill That Can Make You Millions

Chapter 22

Introduction to Technical Analysis

New investors tend to cringe at the thought of learning to understand charts and technical indicators. Some are quick to dismiss it as a hobby for math nerds. If you want to join the ranks of smart money, then **technical analysis** (TA)–the study of charts to determine price trends–should become as familiar to you as speaking in your native tongue. While there is a learning curve, you don't actually need to be good at math, and the time and effort you spend learning the basics will be worth it's weight in gold. In the same way that you don't need to know every last detail in Photoshop to create a digital masterpiece, TA isn't rocket science if you understand the important elements. We will simplify the process in the chapters that follow.

Viewing Charts

Regardless of whether you decide to be a short or long term investor, it is important for you to be able to track the price action for different coins over days, weeks, months, and years. There are a number of sites where you can easily analyze charts on multiple timeframes. CoinGecko, CoinMarketCap, and Messari all have several chart features that are helpful when making comparisons to the broader market. A majority of exchanges also have charts built into their trading panels.

When it comes to honing in on the finer details of a cryptocurrency, smart money favors TradingView, which is specifically designed for reading charts. You can pull up charts for hundreds of coins on over a dozen timeframes, apply trendlines and technical indicators, and even study other traders' ideas and interact using its social media features. Its free plan is more than enough for the average user, and if you want to upgrade to see more timeframes and add more indicators, its paid tiers are very affordable.

TradingView charts have a dizzying array of features, most of which you will likely never use. We are going to narrow it down to the absolute essentials. Spending five to ten minutes each day learning the application is one of the simplest, most effective ways to maximize your investments.

TradingView Simplified

Once you've set up your TradingView account, click the "Chart" tab. Type any letter key, and a search box will appear (See Figure 30). Select the "Crypto" tab, then do a search for bitcoin, or simply type its ticker: BTC. You will notice that BTC trades on many exchanges, and under many different trading pairs. The first step is to search for the pair that you are trading. For example, if you plan to purchase BTC with USD, find the BTC/USD pair. Scroll down the list to find your exchange, which is listed on the far right. Be sure to choose your most frequent exchange, since spreads across different exchanges will create deviations in price. If you click the + symbol to the right of the exchange, it will add the coin to a personalized watchlist panel to the right of the chart. The watchlist is a useful tool for switching between charts.

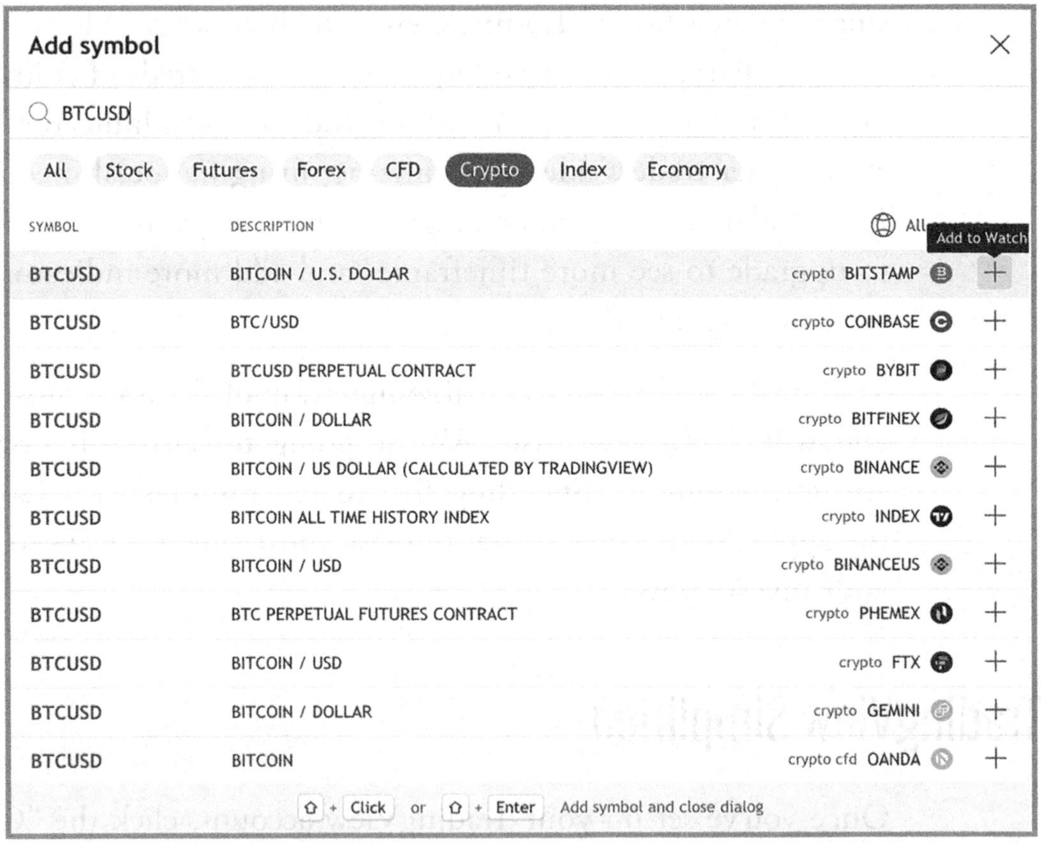

Figure 30: We will use Bitstamp's BTCUSD chart, since it has price data for bitcoin that goes back to 2011.

The most basic charts are line charts, which show the price of an asset as it rises and falls over a period of time. While they can be helpful to spot general trends over months, years, or decades, their displays are limited to closing prices. This means that even if the price of an asset ranges widely on certain days, the only data recorded is the price at market close.

Figure 31: A simple line chart for BTC. Each point reveals the closing price for that day (12:00 UTC). Line charts can be useful for observing general trends over longer time periods.

To see a more comprehensive view, choose the "Candles" option from the chart view icon on the top left panel in your chart. Even though the crypto market doesn't technically close, daily candles close at 12:00 UTC, coordinated universal time. TradingView will automatically display a countdown to the close on the timeframe you choose, which you can find highlighted directly under the current price in the price column. The colored dotted line across the page is the price line, and will point straight to the current market price as it rises and falls.

Figure 32: The default bar chart view of BTC, with a horizontal price line. Notice the current price highlighted at the far right, and beneath that is a countdown to the next candle close (on a 1 day timeframe). Bar charts are the preferred view for investors when making trading decisions.

Candlesticks were first used by Japanese farmers in the 18th century to track the price of rice. Each candlestick is a visual representation of an asset's price over a specific period of time. Candlesticks can be used on any timeframe, ranging from less than a minute to weekly, or even monthly.

You can choose your timeframe from the upper left panel (Figure 33). Candles close according to the timeframe you set on your chart. For example, a 4 hour timeframe will create a new candle six times every day, whereas a weekly timeframe will only create a candle once per week. The timeframe you choose depends on the level of detail you wish to see, which depends on how frequently you trade. For example, day traders who may make multiple trades every day, will use anywhere from 4 hr. candles all the way down to a minute or less. Swing traders will most often use 4 hr., 8 hr., daily, or weekly candles.

Figure 33: TradingView tools are on the left column, timeframes are across the top row, and timeline options are on the bottom. The watchlist panel on the upper right gives quick access to your favorite coins. You can also flag those that are important, create different categories, and reorganize the list by daily changes in price or percentage.

The timelines in the bottom left panel are for expanding your view to fit the screen. For example, if you choose 3 months, the chart will automatically resize from left to right to fit 3 months of price data. The candles will adjust accordingly, in this case showing 1 hour candles.

When viewing a chart for the first time, it is recommended that you start by zooming out with weekly candlesticks, and view the chart over the course of two years or longer. From there you can switch to daily candles, and look at 3-6 month windows.

You can click the star next to any timeframe, as well as any other tool in TradingView, and it will automatically appear in a favorites bar up top. Another helpful feature is the "auto" button in the bottom right panel, which automatically adjusts the chart to fit the page as you scroll from one period to the next. The standard view uses the arithmetic scale, so that the distance between 1 and 2 is the same as the distance between 8 and 9 (Figure 34).

The "log" view (Figure 35) displays prices in logarithmic scale, showing percentage changes in an asset's price. In other words, the distance from 1 to 2 is the same as from 2 to 4, and 4 to 8. This makes it helpful to see a more balanced view over time in order to discover long term trends. The standard arithmetic view is the preferred choice for most investors, but this doesn't make it better!

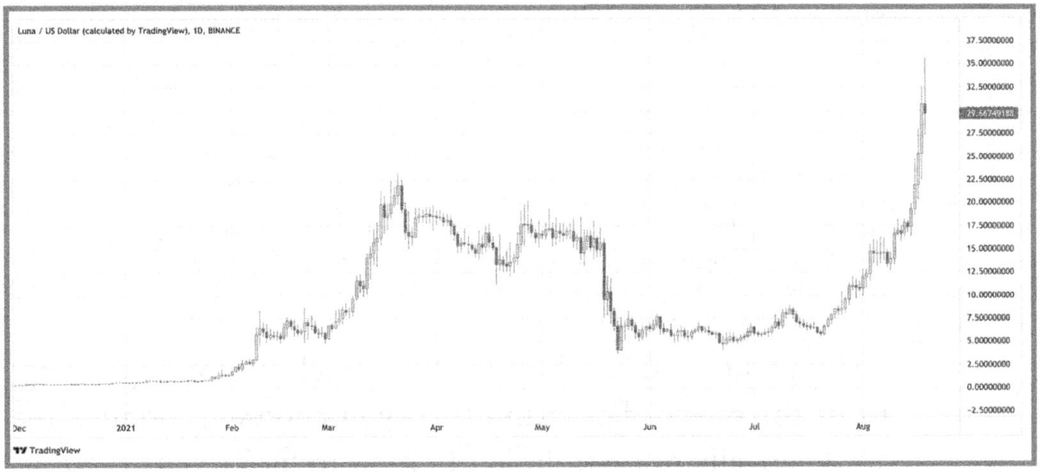

Figure 34: LUNA standard view with the auto feature enabled.

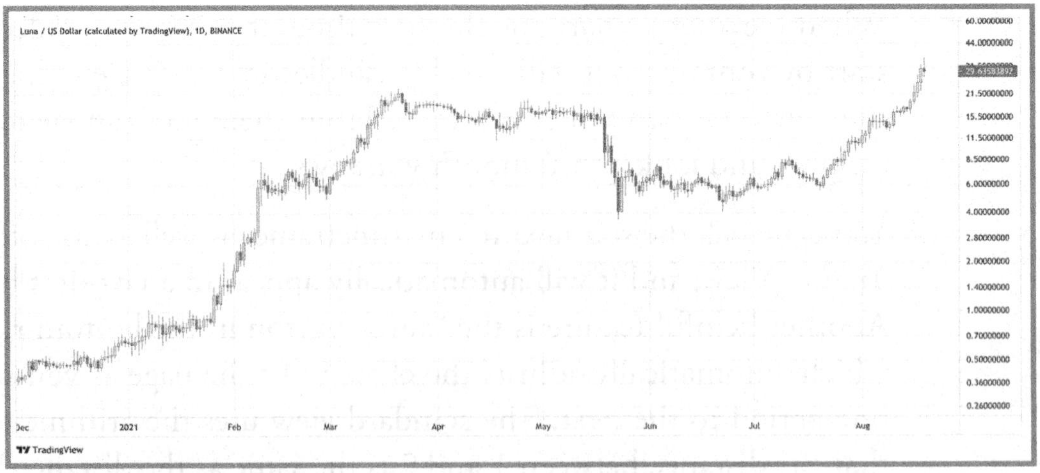

Figure 35: LUNA log view with the auto feature enabled.

Beneath the watchlist you can see price metrics, including the range for the day and year. Another handy feature for more well known cryptocurrencies is a table of key stats with performance for 1 week, 1 month, 3 months, 6 months, year-to-date, and 1 year.

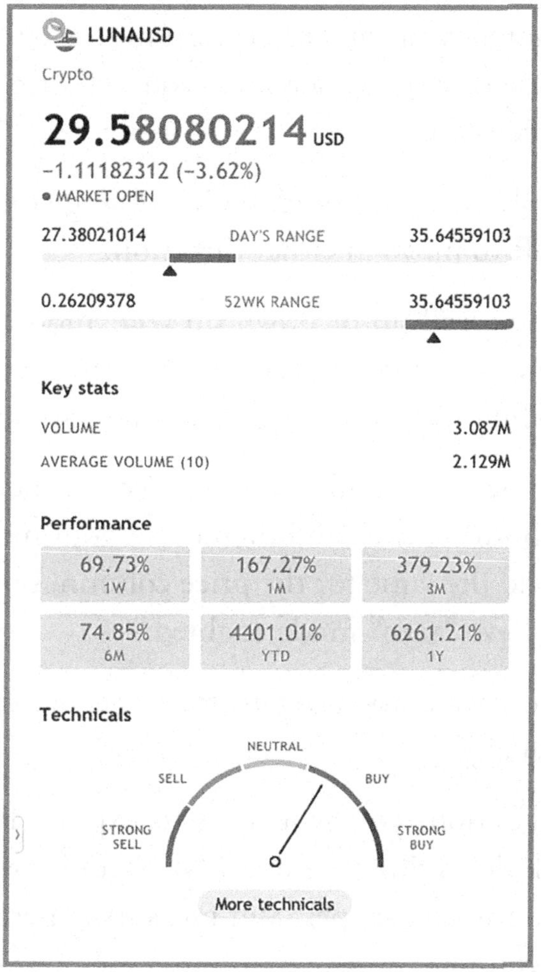

Figure 36: LUNA price metrics.

The simplest way to learn how to use the tools in the left column is to explore one or two options every day. Those that are the most popular, and therefore the most used, are typically at or near the top of each dropdown menu. The crosshairs navigation tool is default, which displays the exact price and date from anywhere on the chart.

Here are 5 navigation tips and tricks to save you hours of frustration when viewing charts:

1. Although you can either use a mouse or a trackpad for navigation, a mouse is much easier since you will often need to combine clicking and scrolling.

2. Simply click and hold while moving your mouse either to the right or left to move in either direction.

3. If you scroll up or down on your mouse, the entire chart view will zoom in or out, respectively. Holding command (⌘ on a PC) while scrolling will zoom relative to one particular point.

4. To quickly expand or contract your timeline, hover over the dates at the bottom, click and then move your mouse either right or left. You can do the same for the price column, but you likely won't need to if you have "auto" mode enabled.

5. If you want to navigate to a specific date, press alt-g to enter it manually.

Helpful actions you will want to explore in the tools column on the left include adding trendlines (alt-t) and text, highlighting areas using a marker, drawing shapes, and measuring percent changes in price over a specified time.

Chapter **23**

Know Your Candles

Playing with Candlesticks

All **candlesticks** have two main sections: a body and a wick, which is also sometimes called a shadow (Figure 37). The body is the rectangle–for a green candlestick, the bottom of the body tells us the opening price, while the top reveals the closing price. Green candlesticks also sometimes appear as white with a dark border. For a red candlestick, the bottom of the body tells us the closing price, while the top shows the opening price (Figure 38).

If you get confused, remember: *green goes up, and red falls down.* Regardless of whether a candlestick is red or green/white, the top and bottom of the wick will reveal the highest and lowest prices during the candle period. This data is equally important to the opening and closing prices of the candle body.

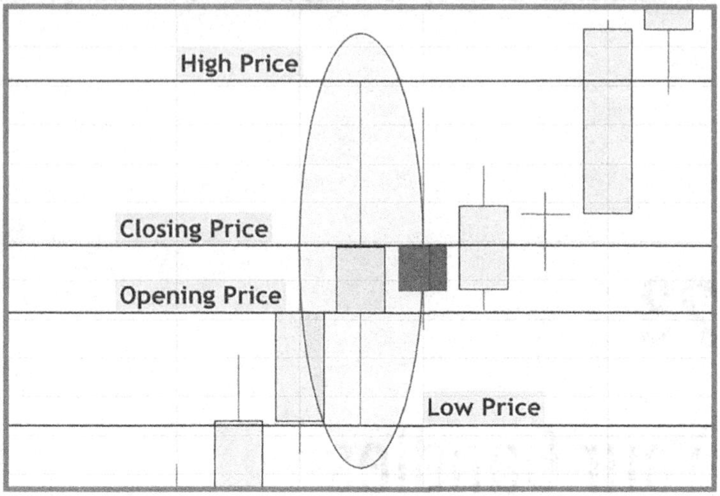

Figure 37: A green candle is formed when the price closes higher than it opened. The price ranges between the upper and lower wicks, with the body of the candle revealing the opening and closing prices.

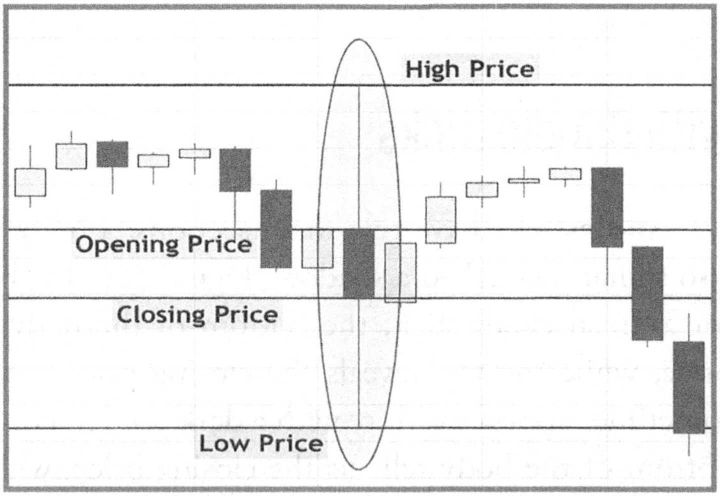

Figure 38: A red candle differs from a green candle in that the price closes lower than it opened.

Some investors use specific candlestick shapes and patterns when determining which trades will have the highest potential reward. There are a handful of candlestick shapes and sizes that can help to reveal either trend continuations, or impending reversals. We will explore price trends and reversals in more detail soon, but for now you should know that *candles are meaningless by themselves. They must always be read within the context of recent market activity.* You should observe at least three fully formed candles before making any judgements. Even then, nothing is certain, since your perspective will be influenced by the timeframe you choose. Still, they can be useful when combined with other methods of analysis. Below we will look at six of the most important candlestick patterns. While you can view these on any timeframe, for mid or long term investors, a daily timeframe is best.

6 Important Candle Patterns

Marubozu candles

Marubozu is Japanese for "bald head." These are candlesticks with little or no wick in either direction. Whether they are green or red, traders aren't likely to bet against candles with strong momentum in one direction.

Figure 39: Marubozu candles are very rare, since there is almost always at least some degree of wick. As with all candles, their meaning is relative to their placement. Here it served as an early warning sign of a significant dip.

Engulfing Bullish/Bearish Candles

These are large green or red candles that completely overshadow the preceding candle. For an engulfing bullish candle, the previous red candle would be completely contained within its body. Another big green candle with even higher volume would be further bullish support.

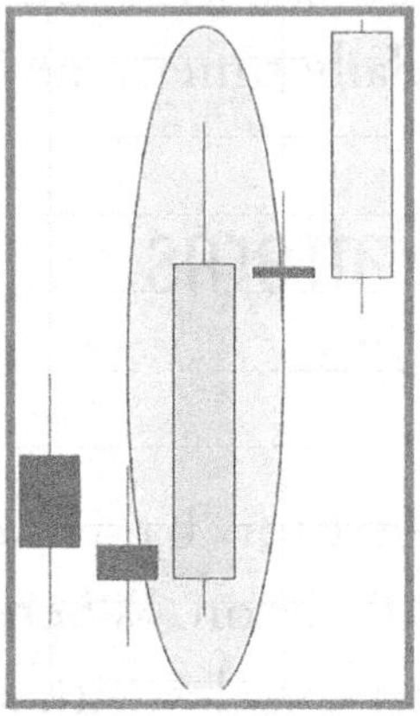

Figure 40: A bullish engulfing candle, which completely eclipses the previous red candle.

Engulfing bearish candles differ in that they are red, and preceded by a smaller green candle. A second large red candle with higher volume strengthens, but doesn't confirm, a bearish case.

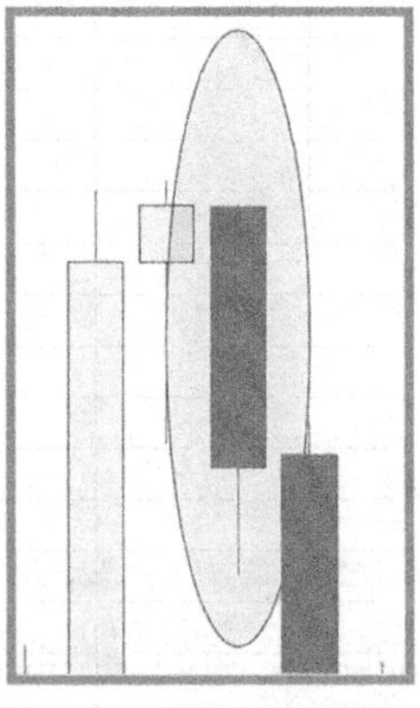

Figure 41: A bearish engulfing candle, which completely eclipses the previous green candle.

Doji

Doji is Japanese for mistake. It happens whenever the opening and closing prices of an asset are nearly identical. The candle has a very small body, with an upper and lower wick of varying lengths. Dojis often look like crosses, and signal a standoff between buyers and sellers.

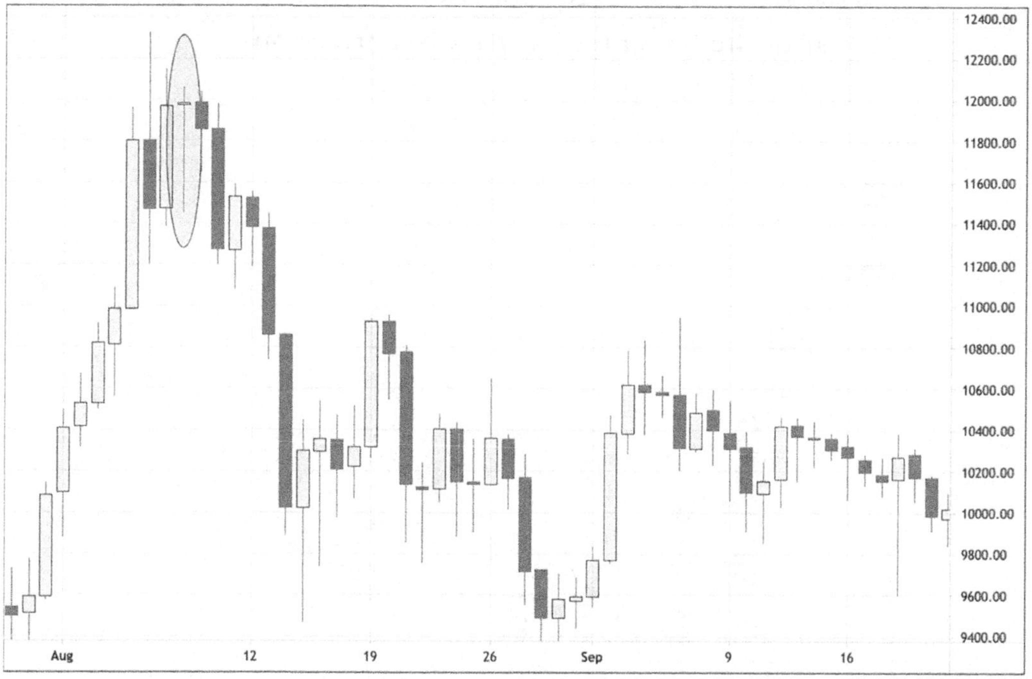

Figure 42: An evening star doji with a very long lower wick reveals heightened uncertainty. Notice the the shift from green to red after the doji's appearance as sellers gain control over buyers.

Dojis can take on several meanings depending on how their neighbors are behaving. For example, if an asset is in a strong uptrend for several weeks, with many large green candles, and suddenly a doji forms, it may be an indication that a reversal is imminent. This type of doji is often referred to as an **evening star**, since it appears after a considerable uptick in price.

Hammer

A hammer appears near the bottom of a downtrend, and has a small body at the top with little or no upper wick. The long lower wick is more than ⅔ the length of the candle's total size (Figure 43). Hammers are bullish signs, since the longer lower wick reveals that sellers have been overcome by a stronger wave of buying. You will often see hammers appear at the tail end of a massive correction.

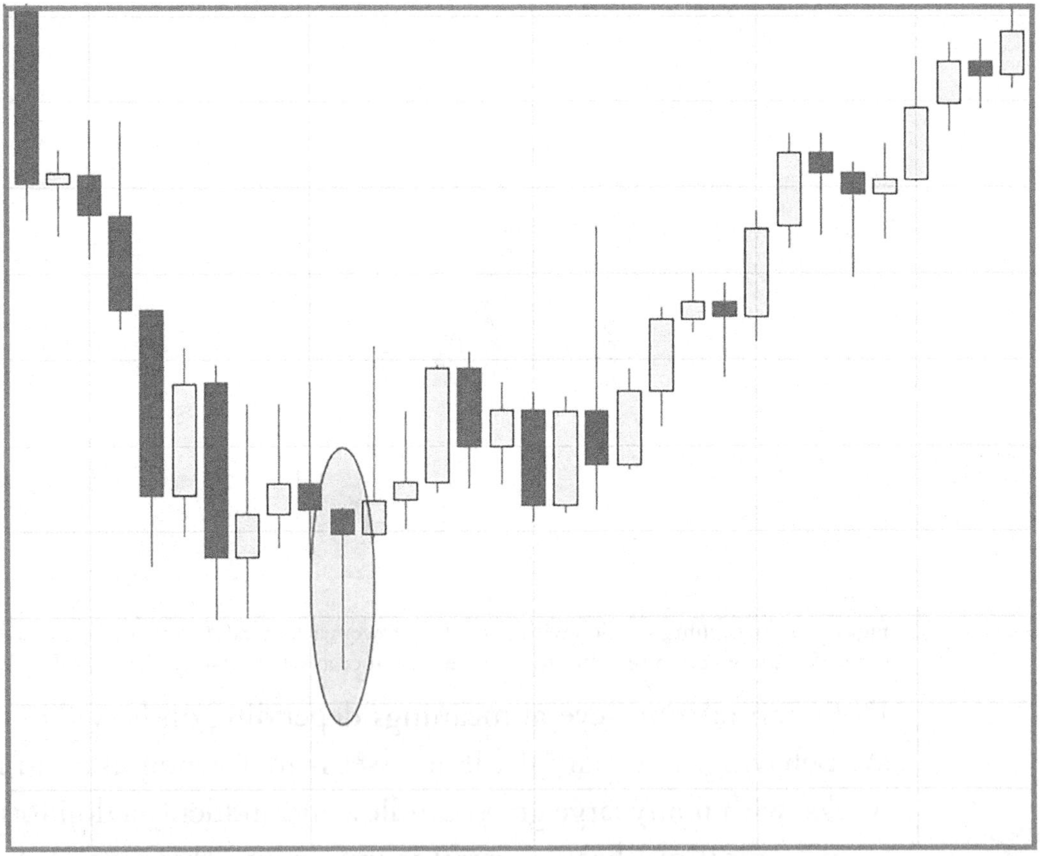

Figure 43: This hammer appears at the bottom of a downtrend, and is one sign of a reversal to the upside.

Hanging Man

A hanging man has the same shape as a hammer, except that it appears during an uptrend (Figure 44). In this case, the long lower wick is interpreted as sellers starting to gain momentum as the price dips temporarily. It is a subtle show of weakness that could lead to a bearish reversal. If the next candle is large and red, it adds to the case for a correction.

Figure 44: This hanging man appears at the top of an uptrend, and is followed by a significant sell-off.

3 Candle Features to Watch

It can be confusing to try and distinguish between so many candle shapes. Your best bet is to learn how to spot the more extreme outliers by closely observing a few obvious candle traits. The first is the overall size, or vertical length of the candle from the lower wick all the way to the top. The larger the candle(s), the more of an effect it will have on the psyche of the collective investor. For example, if a long bodied red candle suddenly shows up amidst a sea of green during an uptrend, it is a red flag that weakens the trend (Figure 45). If, however, a long bodied green candle shows up during an uptrend, then it strengthens the existing trend.

Figure 45: Notice the size of the two highlighted red candles relative to all the other red candles that precede them. Traders who were able to spot these outliers could have sold and been spared from a 60% deeper correction.

The second feature you want to study closely is the length and direction of the wicks themselves. The longer the wick, the higher the volatility (uncertainty). Long upper wicks during uptrends can be considered walls of resistance where sellers were happy to meet the demand. Long lower wicks during downtrends can be viewed as floors of support where buyers quickly absorbed supply (Figure 46).

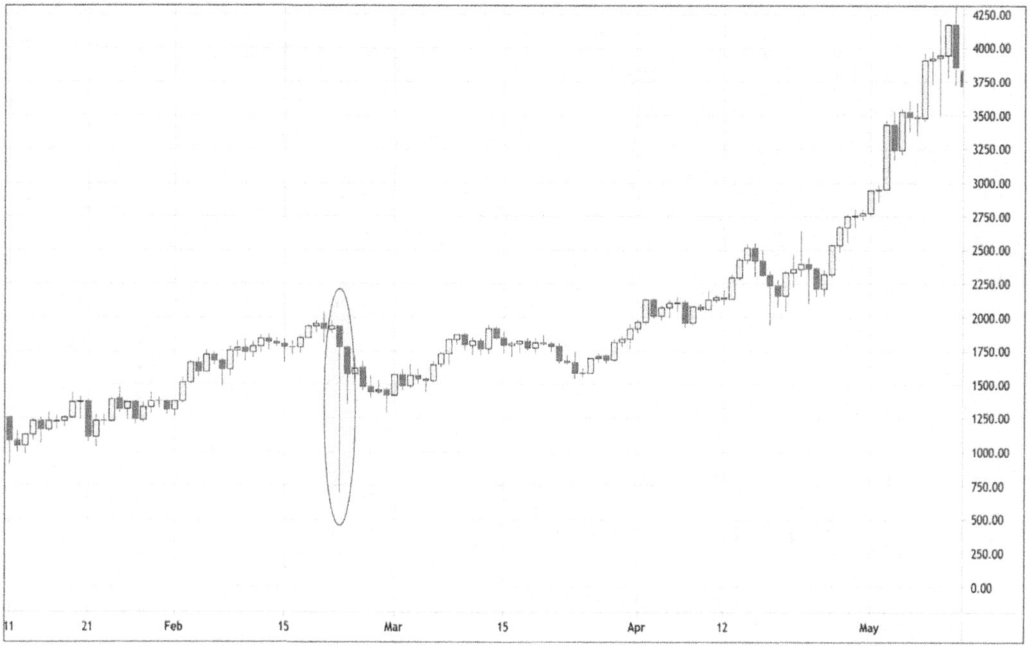

Figure 46: This long lower wick reveals that buyers were eager to absorb supply from sellers, leaving the general uptrend intact.

The third feature to watch is the relationship of larger candles to those that surround it. As an example, if you see a single doji appear where before there were only big green candles, it has a much stronger influence than a doji in a cluster of mixed candles with smaller bodies. *The more a candle stands out in the crowd, the more likely other investors are going to notice and take action.*

Take a look at Figure 47. There is a large bearish engulfing candle that wicks down below four previous green candles. What's more, it appears after several weeks of bullish momentum, where the price increased almost 87%. Since it sticks out like a sore thumb, it is safe to assume that other investors are going to take note, leading to more selling pressure in the days to come.

Figure 47: A bearish engulfing candle after two weeks of mostly green is an early indication of a larger correction.

Be careful not to jump to conclusions as soon as you spot an outlier. Looking at Figure 48, notice the red candle with a long lower wick that appears after seven days of green candles. Even though there was a significant dip that day, there was plenty of demand from buyers to keep the primary uptrend intact. If you had sold at close you would have missed out on an additional 31% gain over the next four days.

Figure 48: A large red candle in a sea of green doesn't always signal an immediate reversal. It may simply be an early warning sign that bulls are beginning to lose momentum.

Volume

Volume is the quantity of an asset that is bought and sold in a specified timeframe. Green candles correspond to green volume bars (more buying than selling), and red candles correspond to red volume bars (more selling than buying). You can view the volume on any chart by clicking the "f_x" icon (or hitting the / key), and typing it in the search box. It will display a bar chart on the bottom that corresponds to the selected timeframe. *Volume is a representation of the emotional intensity of investors.* Take a look at any chart, and you will often see huge volume spikes during volatile moves either to the upside or downside.

Price action should be the most important aspect of chart analysis. Volume can be used to either confirm or reject price movements. If you wait for volume to spike or collapse before buying or selling, many times you will miss out on a big move up or down. Sometimes there is high volume, but the price is consolidating, since the buyer and seller interest is roughly equal. Other times there can be low volume at the start of either an uptrend or a downtrend that gradually increases over several days.

During a healthy uptrend, volume will typically rise as the price increases (Figure 49). Still, don't expect it to be linear! During smaller corrections in an uptrend, volume will drop to show a lack of seller interest, only to rise again as the primary uptrend absorbs the selling.

Figure 49: ADA shows healthy volume during an uptrend, with progressively larger green bars. The large red bar after a sea of green led to a larger correction.

Strong downtrends also often display higher volume, reaching a climax when extreme fear leads to panic selling (Figure 50). **Relief rallies**, when the price temporarily recovers during a downtrend, will be shorter and typically have lower volume. Keep in mind that these are general market trends that won't apply perfectly to every circumstance.

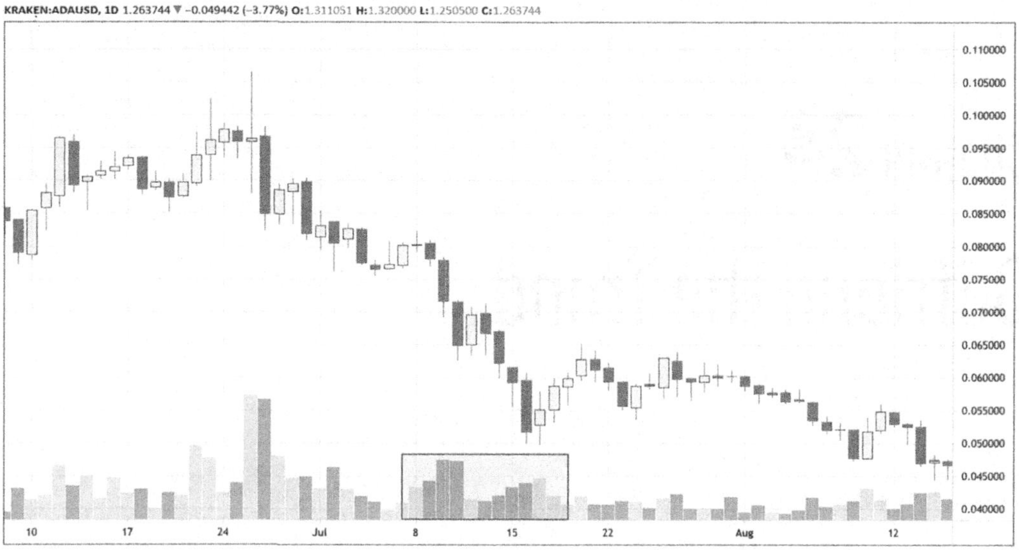

Figure 50: Here ADA shows significant selling volume, which signals a continuation of the downtrend.

Chapter **24**

Befriend the Trend

The Trend Trinity

There are three basic price trends: uptrends, downtrends, and sideways trends. An **uptrend** is characterized by higher highs and higher lows, as the price steadily increases (Figure 51). The opposite of an uptrend is a **downtrend**, which is characterized by both lower highs and lower lows, as the price steadily decreases (Figure 52). **Sideways trends** are consolidation periods where the price is **range bound**: it chops around in a parallel channel with an upper and lower boundary (Figure 53).

For most investors, sideways trends are lazy days, kind of like that introspective emo scene in a movie where the lead actor is walking around with his hands in his pockets, musing over his next move. Incredibly boring, but somehow necessary in the grander scheme of things.

Figure 51: A clear uptrend, with a trendline that connects 3 lower wicks.

Figure 52: A clear downtrend, with a trendline that connects 3 upper wicks.

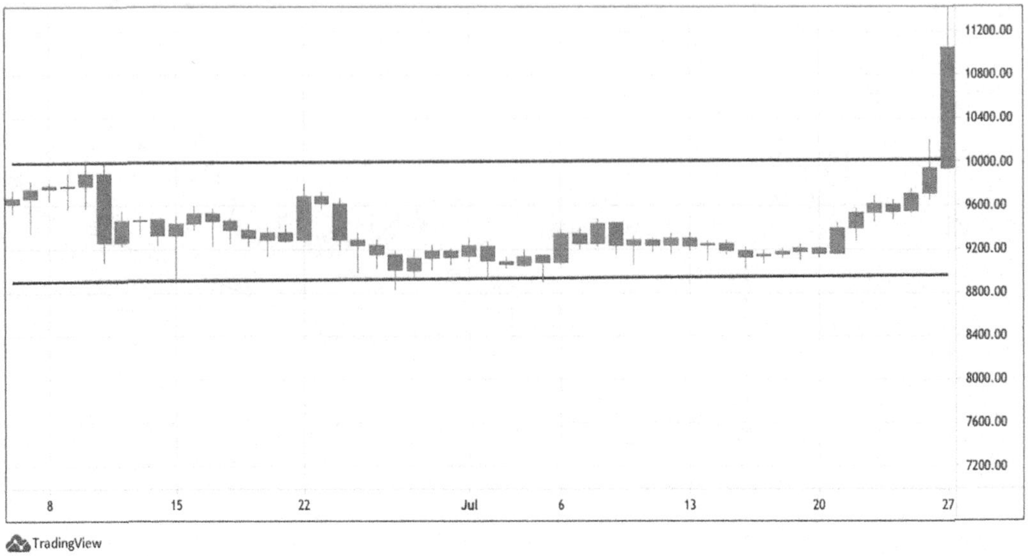

Figure 53: A sideways trend (consolidation phase), where the price remains in a parallel channel.

Spotting Trends

Learning how to spot price trends is a game of connecting wicks. Anyone can learn to do it with relative ease, given enough practice. *The trend is your best friend.* If you can learn to spot and follow the trend, it will reward you one hundred fold. Trend trading is arguably the simplest, most effective trading strategy for the conscious crypto investor.

The easiest way to spot a trend is to use the handy trendline tool in TradingView (alt-t). Begin by viewing a 3-6 month time period. To create a **trendline**, all you need to do is draw a line that connects the wicks of at least three candles (Figure 54). For an uptrend, you will want to connect the lower wicks, while in a downtrend you will want to connect the upper wicks. Extend the line as far as you can on the chart.

Candles are often connected by their wicks, since there is almost always at least some wick on either end of a candle. Note that it won't always be practical to connect the tips of the wicks. It is sometimes OK to cross through elongated wicks when drawing trendlines. But if you need to cut through a candle body, then it's not a trend!

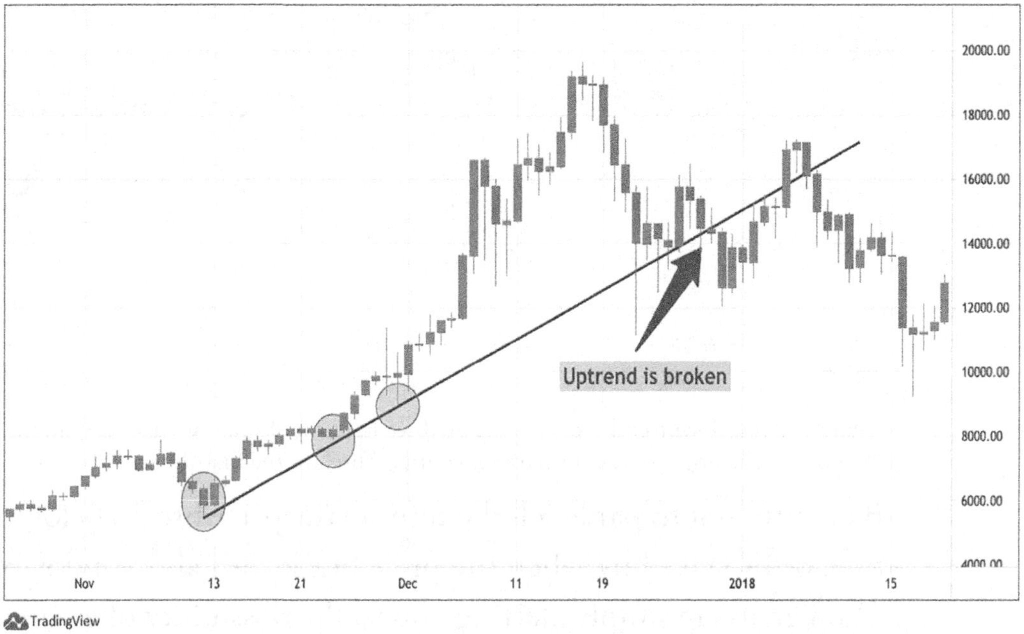

Figure 54: This uptrend is established by connecting at least three lower wicks. The trend is broken, or invalidated after a daily candle closes below the trendline. Even though the price briefly pulls back above the trendline, the uptrend can no longer be considered intact. It is soon followed by a reversal.

The reason we connect the lower wicks during an uptrend is to establish clear levels of support. When the trendline is broken, it shows that support has failed, which signals a reversal to the downside. Similarly, we connect upper wicks during a downtrend to establish levels of resistance (Figure 55). When the trendline is broken, it signals a potential reversal to the upside. *The longer the trendline, the higher the likelihood of a trend reversal after it breaks.*

Figure 55: This downtrend is connected by three upper wicks that span several months. After almost 10 months, the price breaks above the downtrend line for the first time.

Be careful not to panic sell the minute the price breaks below an uptrend line, or FOMO buy when the price breaks out above a downtrend line. The signals are simply alerting you to the possibility of a reversal. Expect the occasional **fakeout**, when the price temporarily breaks above or below a trendline, then suddenly pulls back to resume its normal course. This is why smart money uses **confluence**: a comprehensive trading strategy that uses a combination of multiple signals.

As an example, let's add volume bars to the uptrend from earlier (Figure 56). Notice the large red candle that pierces the trendline. Not only is there a large red candle revealing a significant drop in price, but there is accompanying high volume. These confluent factors give more weight to the argument for a trend reversal.

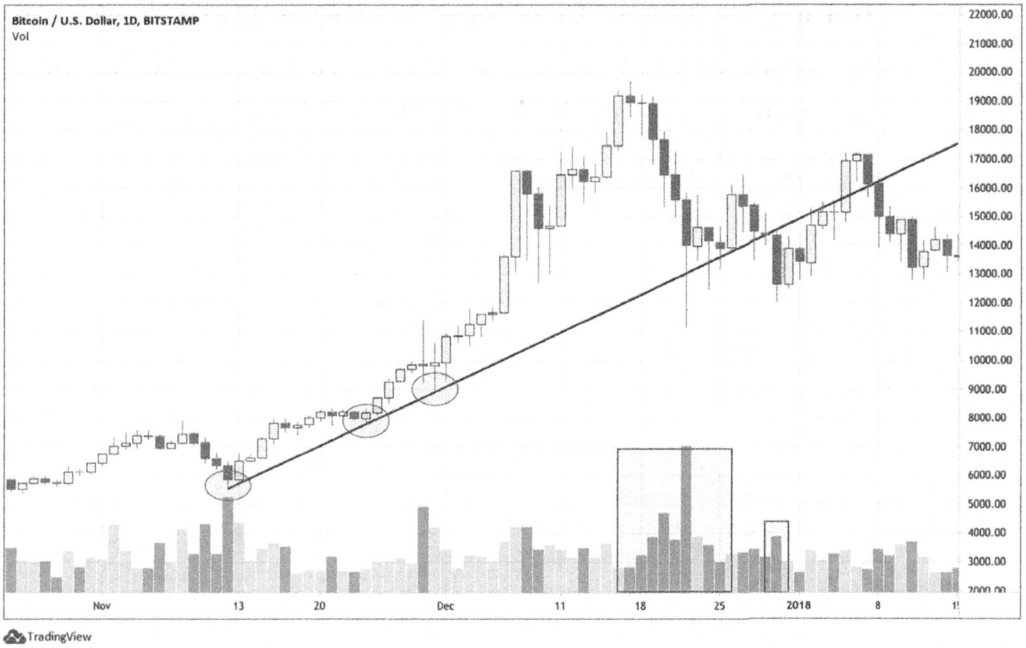

Figure 56: There is heavy volume both leading up to the first dip below the trendline, and on the first day that closed beneath it. Volume increased during the sell-off, which strengthened the case for a reversal.

Signs of an impending trend reversal become clearer on longer timeframes. A 30 minute or 1 hour chart could reveal several false signals as the price approaches a trendline. On the other hand, a daily, or even weekly candle that closes below a trendline is a much stronger signal, since it encompasses more information. *The longer the timeframe, the more reliable the trading signal.*

Trading Channels

If you are able to draw parallel trendlines that connect both upper and lower wicks during any type of trend, you will have created a **trading**

channel, which can be helpful to identify possible price ranges within the trend. In properly drawn channels, you will often notice that the price will bounce off the boundaries with more force.

Figure 57: This trading channel for LINK is gently sloping upward during a clear uptrend.

Trading channels are rarely nice and neat, as can be seen in Figure 57. Even though the price pierces both the upper and lower boundaries, a significant amount of price action stays inside the channel. Instead of ignoring the channel in favor of these outliers, a better strategy is to mentally visualize **buffer zones**: a 1-5% cushion for the price to move either above or below a trading channel without invalidating it. Buffer zones provide some flexibility in the event that a departure from a trading channel is only temporary.

Support and Resistance

Horizontal support and resistance lines are also examples of trendlines (Figure 58). **Support** can be seen when the price of a coin drops to a certain area and then bounces up and away two or more times. **Resistance**, on the other hand, can be seen when the price of a coin rises to the same area and then drops back down again two or more times. *Support is the floor, and resistance is the ceiling.*

Figure 58: Support and resistance levels for AAVE. Notice how even after clearly breaking out of the channel, the price repeatedly bounces off the support floor during a major correction. The longer the price trades in a range, the stronger the support and resistance become.

On some charts, you can draw support and resistance lines that span many months or years. It may seem like a resistance level from three years ago is insignificant, but take a look at the chart for DASH in Figure 59.

Think of all the people who may have taken positions at around $408 during the last bull market who have been patiently waiting for it to return to sell and break even. Wolves who anticipate this scenario add to the selling pressure by **front-running** the market, which in crypto simply refers to making moves before everyone else. In this case, they front-run by setting sell targets at this zone.

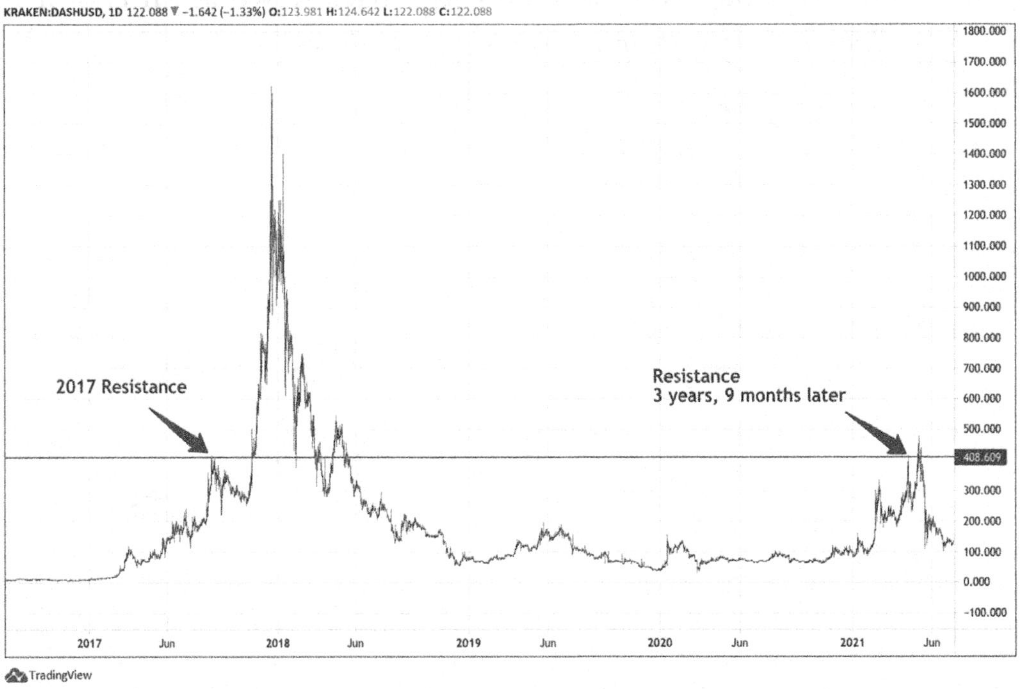

Figure 59: Price has memory that can span many years. When DASH reached the resistance level shown for the second time, it led to an almost immediate 45% dip. Smart money was clearly selling at resistance and buying the dip before the next leg up.

The strength of a support or resistance line can be gauged by how quickly the price bounces from it. If you see a strong bounce up and away from a support line accompanied by higher volume than usual, it is a sign of relative strength. Just like with trading channels, it is helpful to create a 1-5% buffer zone with your support/resistance line in the middle (Figure 60). You can use the rectangle tool to draw support and resistance zones. This will keep you from panic selling when a support line is temporarily broken, or FOMO buying as soon as a resistance line is crossed.

Figure 60: Adding buffer zones helps to account for slight deviations from support and resistance lines. Here you can see significantly more price action is contained within the buffer zones.

On shorter timelines, support and resistance are weakened the more they are tested. Think of a jackhammer that is repeatedly hitting a floor. The floor loses its integrity the more it is struck, until finally caving in to a **breakdown**: price movements down and away from a key level of support. On the other hand, if you notice that an upper resistance zone has been tested several times in a few days, there is a higher likelihood of a **breakout**: price movements up and away from a key level of resistance.

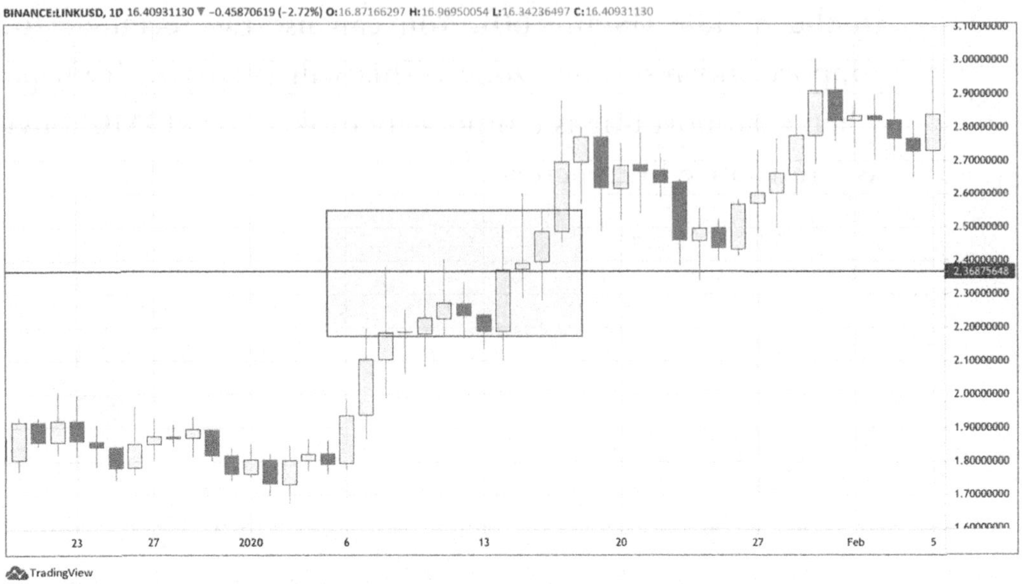

Figure 61: This resistance line was tested four times in one week, leading to a breakout after the fourth occurrence. Notice that resistance later flips into support when the price temporarily corrects before making its next move up.

Using Trends to Choose Your Investments

When viewing the chart of a potential investment, you want to look for smoother waves that form over several weeks or months. To be sure, ask yourself the following questions:

1. Do the price waves reveal clear trends that are easy to label? You should be able to draw multiple trendlines either up or down that connect three or more points, and also be able to draw parallel channels that reveal healthy phases of consolidation (Figure 62).

2. How long do the trends typically last? There is a big difference between a trend that lasts for a week or two vs. several months. Longer trends reveal consistency, and at the same time allow you to more easily spot

signs of a reversal. This makes it easier to enter and exit at the right time.

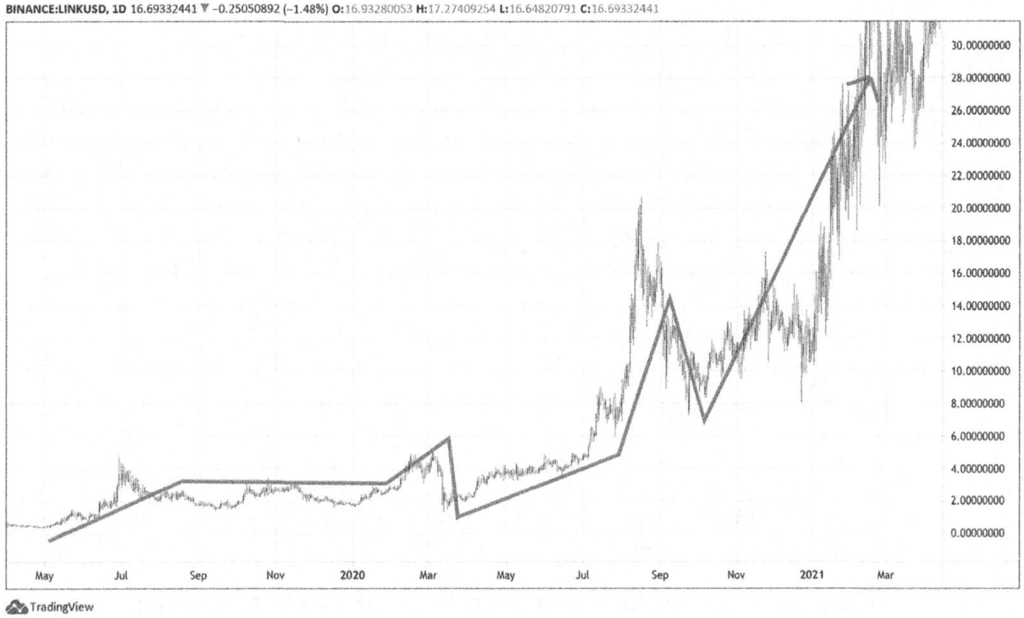

Figure 62: Notice the trend consistency for LINK over several years, as indicated by the solid line. There are clear uptrends, sideways trends, and downtrends. For this time period, LINK was either trending up or sideways for much longer than it was trending down. From a TA perspective, LINK's stability would make it a prime candidate for investment among long term investors.

Thrill Seeking Charts

Coins whose charts are characterized by sharp, choppy waves can definitely be an exciting ride! We will refer to these coins as **oscillators**, since their prices tend to move in an unpredictable way similar to roller coasters (Figure 63). Sometimes they consolidate in a range for weeks, months, or years, only to spike up dramatically in a couple of days before eventually crashing back down again. Oscillators exist in every cryptocurrency category.

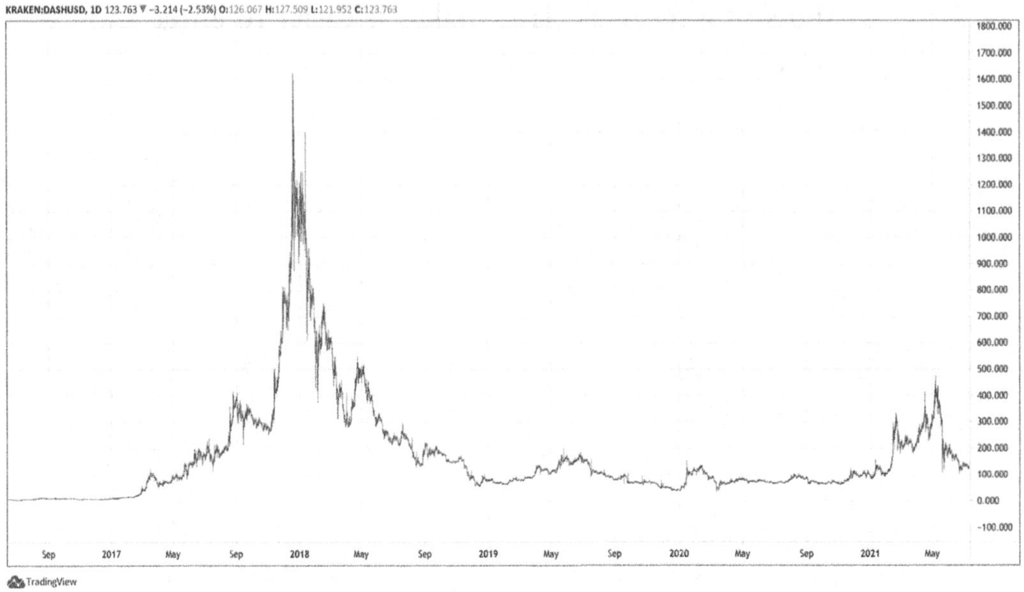

Figure 63: DASH is a prime example of an oscillator, with extreme price swings that can happen rapidly and unexpectedly.

Most investors who are drawn to oscillators get a thrill out of the prospect of their investments exponentially increasing in a short time period. Day and swing traders who like to gamble are drawn to oscillators like dogs to peanut butter. But how many of them end up profiting from these investments? With short and infrequent upswings, most of the time the coin is either trading sideways or taking a dive. *The higher the volatility of a coin, the lower the likelihood that you will be able to properly time your entry and exit.*

Oscillators are the most unstable type of cryptocurrency. Most who are invested in them have a short term outlook with the motive to flip their holdings for a quick buck if and when the price moons. As a long term investor, would you rather be navigating relatively smooth waves, or choppy and unstable ones? You definitely don't want to be stuck holding the bag when the whales who bought theirs three years ago dump it all at once!

Vaporware

When a crypto project has been classified as **vaporware** by the disgruntled CT gods, it means that the technology it proposes is still only a concept in development. While they may have mouth-watering marketing and grand visions for the future, they haven't yet released a working product with a strong network effect. This makes their value even more speculative than their crypto relatives, which generally causes them to dip much lower during larger corrections.

Not all projects that are labeled as vaporware are bad investments. Cardano (ADA) is one such example, since it took seven years for it to enable Dapps to be built on its network. At the same time, its chart shows strength and stability. During corrections it has held its value better than nearly every other altcoin. In fact, in 2020 its price rose 400% more than bitcoin.

Cardano also has a clear roadmap and an enormous social community, with a large majority of its investors staking their coins on the network. Their success will ultimately be determined by how well they deliver on their promises.

Figure 64: Cardano's chart speaks for itself. ADA was one of the top performing altcoins in 2020.

Chapter 25

Moving Averages

A **moving average** is an asset's average price over a given period of time, which is used to help recognize and clarify existing trends. The 20, 50, 100, and 200 day simple moving averages (SMA) are frequently used by traders, which are calculated by adding up the closing prices for the time interval and dividing the sum by the respective number of days. They are represented on charts as smooth lines that either slope up or down, changing as newer price data is factored into the equation.

Exponential moving averages (EMAs) give more weight to recent price data, and are therefore considered to have less lag than simple moving averages. The figures in this chapter all use EMAs. In TradingView, type "EMA" into the indicator search box, and you will see a number of options. In this book we are using the "EMA 20/50/100/200" by drsweets. You can also customize the length of a standard EMA indicator by clicking the settings icon and selecting "Inputs."

During a clear uptrend, moving averages will slope upward, and will be stacked one on top of the other with shorter MAs either on or nearest to the top (Figure 65). During a clear downtrend, moving averages will slope downwards, and will be stacked one on top of the other with shorter MAs on or nearest to the bottom (Figure 66).

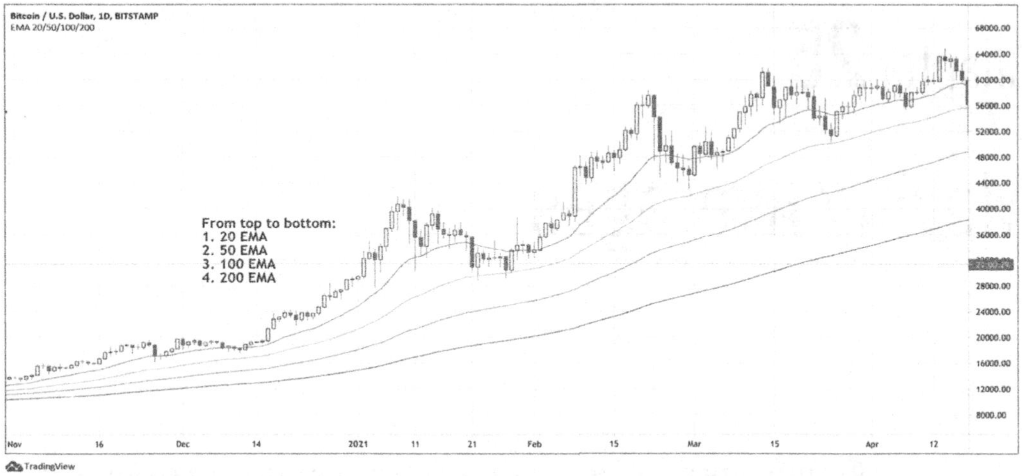

Figure 65: All four EMAs are sloping upwards, with shorter MAs on top and no sign of crossover, which are all indicative of a healthy uptrend.

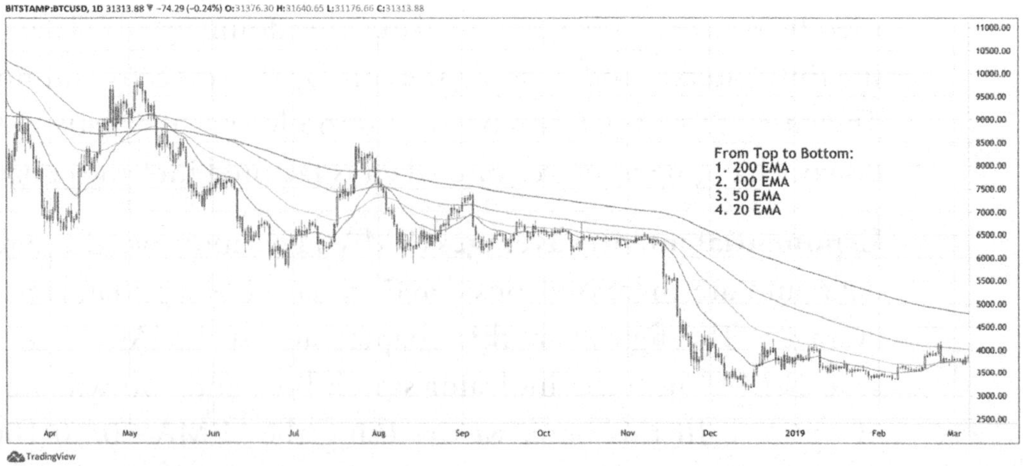

Figure 66: All four EMAs are sloping downwards, with longer MAs on top and some crossover, indicating a strong downtrend with some indecision in the earlier stages.

It is also possible to view weekly moving averages. If you search for the "Bull Market Support Band" indicator, it will display a band between the 20 week SMA and 21 week EMA. Make sure that your timeframe is also set to 1 week, or the indicator won't be accurate. When prices are above both averages, the bulls are in control, and when prices fall below, bears are making mischief.

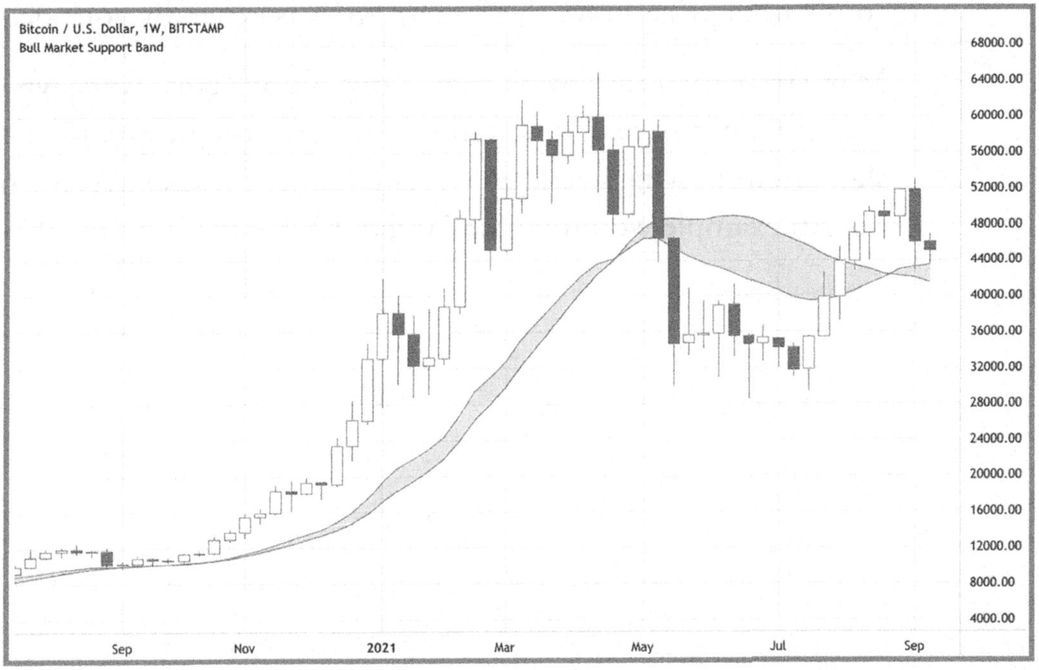

Figure 67: The Bitcoin Bull Market Support band on a weekly timeframe includes both the 21 week EMA and the 20 week SMA. Prices above the band support a bull market, while prices below signal a bear market.

Moving averages are considered **lagging indicators,** since they use price data that has already been recorded. While they aren't meant to predict future prices, they are useful as visual representations of existing trends. Many traders pay close attention to moving averages, which creates a collective consensus that reinforces their importance. Basically, the more people who believe in the significance of an indicator, the more important it becomes to the overall market structure.

The 50 day SMA is an example of collective consensus, since it is watched by countless traders. It is used as a benchmark for support in an uptrend, and resistance in a downtrend. When the price falls to the 50 day SMA in a strong uptrend, supply is quickly absorbed by buyers, and the price increases. When the price rises to the 50 day SMA in a strong downtrend, supply overwhelms demand, and the price drops. A coin with a price above the 50 day SMA is generally considered bullish.

Many traders pay attention to moving average crossovers, where moving averages on different timeframes cross. A **golden cross** occurs when a short term moving average crosses a longer moving average on the way up. An example would be the 50 day EMA crossing the 200 day EMA.

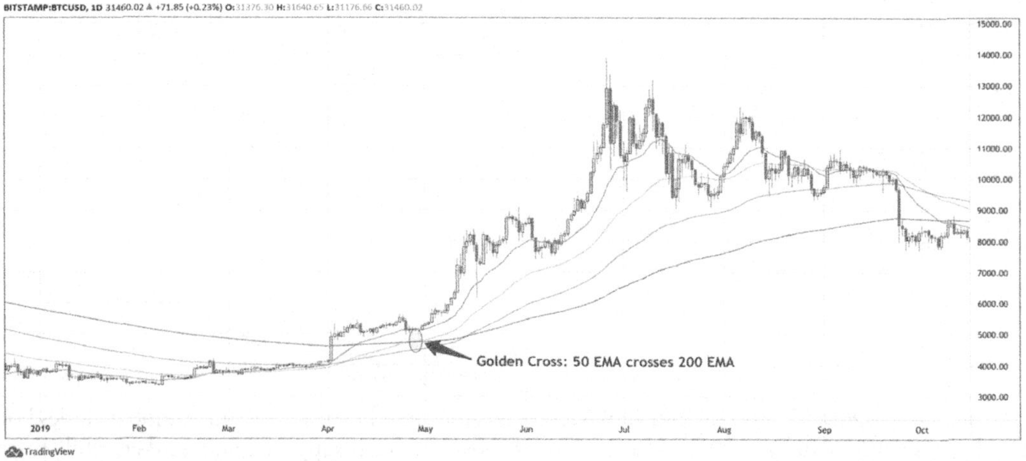

Figure 68: A 50/200 EMA golden cross that confirmed the end of the 2018-19 bear market. Even though the price had already been on the rise for several weeks, using longer time periods gave more strength to the signal.

In contrast, a **death cross** occurs when a short term moving average crosses a longer moving average on the way down. This signals a possible reversal to the downside. While you can technically use any two MAs for crossovers, generally the longer the gap between the two, the stronger the signal.

Figure 69: A 20/50 EMA death cross after the 2017 market cycle top. Notice how the price was already declining well before the death cross, despite using shorter moving averages. MA crossovers should be used to confirm existing trends, not as trading signals.

There are two main drawbacks to relying on moving average crossovers to signal trend reversals. First, they will appear differently depending on the timeframe, with more crossovers on shorter timeframes. Second, crossovers are derived from lagging indicators, and therefore aren't intended to predict future price movements. Similar to moving averages themselves, they are useful for recognizing underlying trends and periods of indecision. For a trend trader, when the moving averages aren't clearly sloping upwards, it is a sign to have as much dry powder as possible.

Chapter 26

Corrections: An Enemy or a Friend?

Whenever the price of a coin drops 10% or more, it is considered to be a **correction**. Corrections can range from a few hours, to days, weeks, months, or years. Assuming you are invested in a fundamentally strong project with actual utility, corrections are healthy recalibrations, readjusting the price of temporarily overvalued coins. They also provide excellent opportunities to buy the coin on sale.

Figure 70: A typical correction during an uptrend can range anywhere from 10-35%. These are both healthy and necessary for the growth of a cryptocurrency.

Think of corrections as wounds. Some wounds are scratches and bruises that are only surface deep. An example of a scratch in crypto would be a coin that's price increases 30%, followed by a 10% drop before continuing its uptrend. Scratches can be shrugged off in hours or days (Figure 71).

Next you have cuts that penetrate through the surface and cause some bleeding. They need to be patched, and it may take some time for the bleeding to stop. Larger corrections of 15-25% that last more than a few hours are cuts.

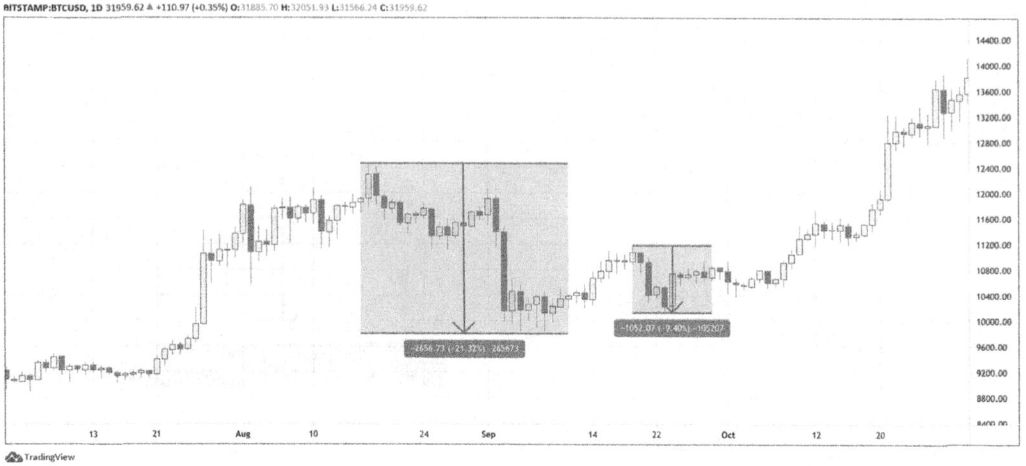

Figure 71: A scratch correction of ~10% on the right, which follows a deeper cut correction of ~21% on the left.

Finally there are stab wounds, which are considered **systematic corrections** when they are marketwide. These are debilitating wounds that will have aftershocks for some time after the initial trauma has passed. Depending on their severity, this type of correction can take weeks, months, or sometimes even years to fully heal (they may never recover!). These would include corrections of 30-90% where the coin either experiences a sudden drop, and/or "bleeds out" with the price continuing to drop steadily over time.

An example of a systematic correction would be what followed the 2017 **blow off the top** event: a sharp drop in the price of a cryptocurrency (in this case bitcoin) after a parabolic rise. A blow off the top signals the beginning of a bear market where the entire cryptosphere experiences a slow, painful bleed out.

Looking at Figure 72, you can see a 43% dip that happened from December 17-22, 2017. This was followed by what is known as a **bull trap**: a temporary relief rally where bulls buy the dip thinking the price will recover relatively quickly, only to be overcome by bears driving the price farther down. From December 22 to January 6, the price temporarily recovered 55% before dropping another 65%. This was the first in a series of aftershocks that lasted until December 16, 2018.

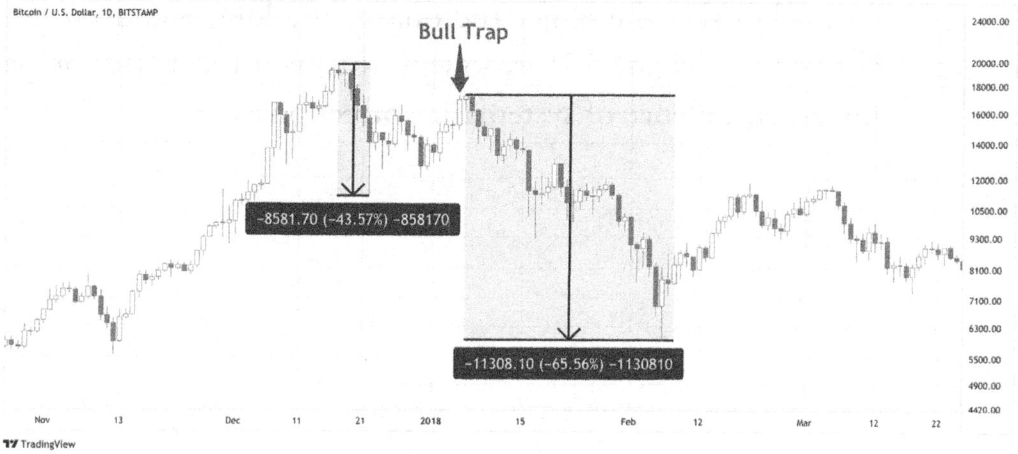

Figure 72: Two stab wound corrections after the 2017 blow off the top, with a bull trap wedged in between. These led to a systematic correction for the entire cryptosphere, and triggered the 2018-19 bear market.

Systematic corrections have more of an impact on people's emotions. There is a saying in the markets: "If they don't scare you out, they'll wear you out." Wolves will sell large quantities of an asset at its peak to drive the price down rapidly. This scares many less experienced sheep into panic selling. Others hold on during the initial drawdown, expecting the price to recover. They are hopeful when it begins its **retracement** as the price moves back up toward previous highs, but this is again followed by fear when it drops even lower than before. Their hope slowly fades over time with repeated aftershocks that continue to drive the price down lower than their entry. **Capitulation** occurs when the last of the **weak hands**–investors who are susceptible to selling their bags after dramatic price movements–finally give in to the pressure and sell, and the price bottoms out. It can usually be identified as a long lower wick in a downtrend.

Corrections are rarely linear, and the deeper the wound, the longer it will take to heal. Corrections are very similar to seismic readings after an earthquake: the larger the quake, the longer and more intense the aftershocks (Figure 73). Knowing this, you can better prepare yourself for the turbulence of systematic corrections.

Figure 73: After the stab wound correction in 2017, it took 510 days for the aftershocks to wear off. The downtrend was finally broken in May 2019.

Ironically, the moment the price reaches rock bottom is precisely when wolves (smart money) sneak back in to buy. The cycle begins anew, with another wave of naive sheep (dumb money) entering the market during the next euphoric rise in price. These market fluctuations will exist as long as the collective human psychology remains the same. Since we have cycled from fear to greed and everything in between for thousands of years, there's a pretty good chance this time around is no different.

You will often hear many people say that everything has changed with so many institutions pouring into the cryptosphere. They will tell you that we are in the midst of a **super cycle** that will last much longer than previous bull runs. But remember that institutions are merely representatives for large groups of people. It's no different from believing we can somehow cheat death if we pretend it doesn't exist, even though people are dying around us every day.

Many who have experienced previous bear markets, which have historically lasted 1-2 years on average, are obsessed with figuring out exactly when the current bull run will end. They count the precise number of days of previous bull and bear markets, track what the price did since each halving, and use sophisticated indicators to try and predict the future.

There are other, more obvious signs for those who are paying attention. At the peak of the 2017 bull run from November 12 - December 17, the price of bitcoin rose 3.5x. In five short weeks, the price and momentum skyrocketed. When you see a steep incline in the price of *any* coin that is sustained for several days or weeks, you are witnessing a parabolic price movement. In other words, that rollercoaster ride you are on is inching toward that moment just before free fall. This is what is meant by "the bull climbs the stairs, and the bear falls out the window."

The average investor shudders at the mere thought of a correction. They treat them like door-to-door salesmen at dinnertime. Denial, regret, and a whole host of other negative emotions surface when the inevitable happens: the clever dip skips the formality of knocking, walks in and sits down to eat your plate!

The bad news is that it isn't possible to avoid corrections altogether. But if you learn to receive them as proper guests that are part of a larger market cycle, then corrections can be your best friend.

Chapter 27

Market Cycles

There are four major stages of every **market cycle**: accumulation, markup, distribution, and decline. The ability to recognize their unique patterns is absolutely vital to the success of any investor. If you don't know which stage the market fits into, you will be behind the curve when it comes to buying and selling crypto.

Accumulation

Accumulation is the first stage, which is the transition from a bear market into a more neutral sideways trend. Visually, accumulation is like a faint, yet steady pulse on a heart rate monitor. It appears after a crash that sends the price down in shockwaves for weeks or months. The asset is trading sideways, with the price appearing to bottom out–an early sign of convalescence.

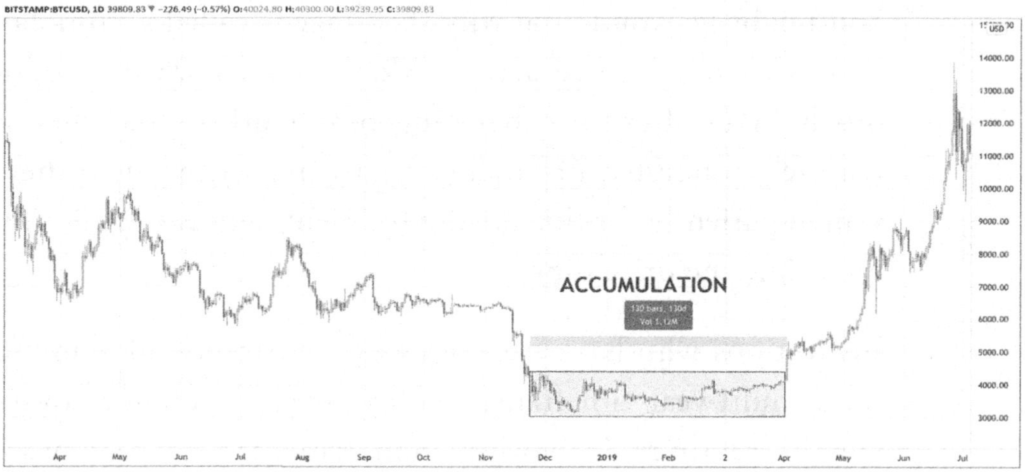

Figure 74: This accumulation stage at the tail end of the 2018-19 bear market lasted over four months, during which the price was range bound.

During the initial phase of accumulation, moving averages continue to slope downward. They start to flatten out, and will eventually cross as the sideways trend matures. Higher lows may form, but higher highs are still only a distant memory.

Figure 75: A closeup of Figure 74, this time including moving averages. Notice the gradual decline in slope of each EMA. They began to cross as an early indication of the pending breakout to the upside in April. The price began to trade in a tighter range as the breakout approached.

You might be wondering why this stage is called accumulation, when there isn't much price action. Most retailers are still coping with so much FUD after an exhausting bear market, that they aren't even remotely interested in buying. Smart money, on the other hand, has waited patiently for this precise moment, and is buying in increments each time the price dips.

Institutions who have seen signs of a bottom, and who are investing on a multi-year timeframe, see the reward potential overshadowing any short term risk. They buy in increments because they are investing upwards of hundreds of millions, even billions of dollars. To buy in one fell swoop would wipe out liquidity on the exchanges, which means they would end up paying much higher prices. Accumulation reveals that buyers are slowly, steadily gaining control over sellers. This is an early indication of a trend reversal as the market cycle transitions into markup.

Markup

Markup is the second stage of a market cycle, which begins when a coin's price makes a clean breakout from its accumulation channel. An uptrend becomes apparent, indicated by higher highs, higher lows, and a steady increase in buying volume. Moving averages will also begin to slope sharply upward. The steeper the slope, and the further the MAs are from one another, the stronger the uptrend.

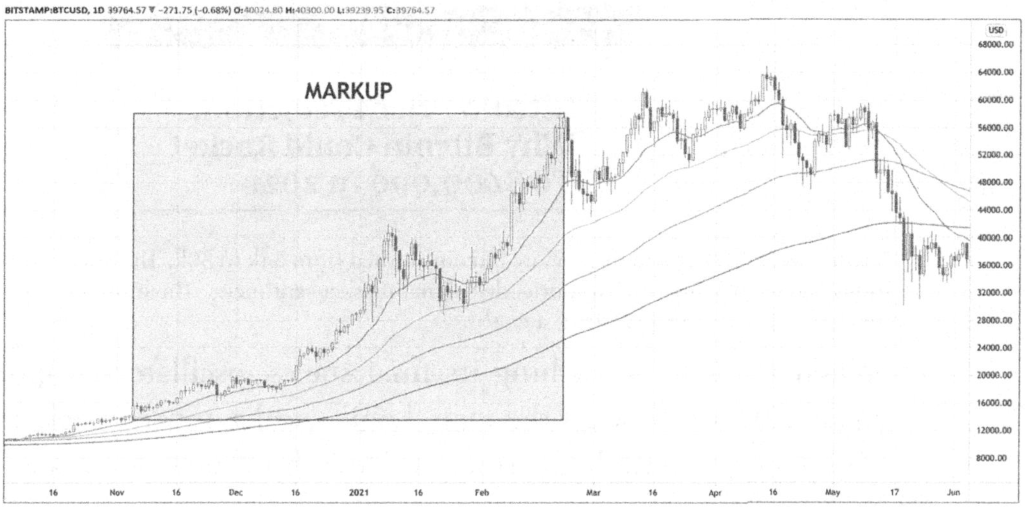

Figure 76: Markup, when the green monster throws a raging party, and nearly everyone gets intoxicated with greed. Buyers beware!

During markup, there may be several short term pullbacks, where the price corrects anywhere from 5-40%. These are healthy pauses in the uptrend, and as long as the price doesn't dip below the previous low, you can generally assume that the uptrend will continue. *Markup always gives rallies the benefit of the doubt.* In other words, pullbacks are tasty snacks for the raging bull as it makes its climb–corrections are easily digested and left behind.

Markup is also when the general public begins to renew their interest in crypto. Hope evolves into optimism and ultimately greed, which leads to increased media attention. This is when it is especially important to cast out deception and false beliefs that the market will gracefully usher everyone up the stairway to lamboland.

Forbes

Apr 8, 2021, 07:50pm EDT | 217,336 views

Bitcoin Price Prediction: Why Bitcoin Could Rocket To $400,000 In 2021

Figure 77: Published a month before bitcoin crashed from 64k to 30k. The only value of price predictions is that they indicate euphoric, delusional market sentiment. The more price predictions begin to circulate, the more imminent a crash.

When markup is reaching its final stages, oscillators, vaporware, and meme coins will be all the rage. Logic sneaks out the back door as even obscure coins double or triple in a few short days or weeks. Now is not the time for pride or arrogance. Now is the time to study the charts closely for signs of weakness. It may appear as a deeper pullback on higher volume that creates a lower low. It may be a slower retracement after a pullback. You may see a **bearish divergence**, where the price is moving up as either volume or momentum is decreasing. All of these are red flags signaling a transition into distribution.

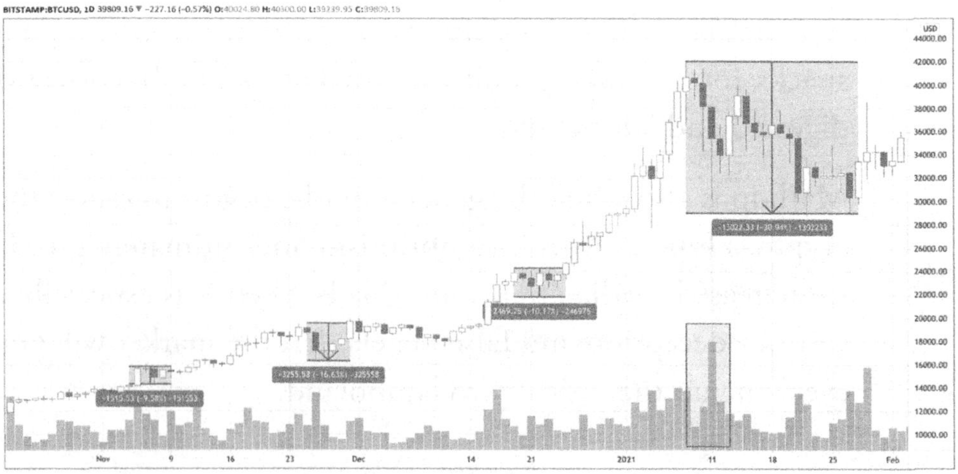

Figure 78: As markup progresses, corrections evolve from scratches to stab wounds. Notice the depth of the final correction, the accompanying volume, and the time it took to recover compared to the three previous corrections. These are all red flags that the green monster is about to fall flat on its face.

Distribution

The third stage of a market cycle, known as **distribution**, begins as bulls and bears enter a standoff similar to accumulation. In the earliest phase of distribution, prices may temporarily continue to climb higher despite decreasing volume and price momentum. This tricks many retailers into complacency, falsely believing prices will soon soar to new heights.

Figure 79: Distribution, the third stage of a market cycle.

Moving averages will start to lose their slope during distribution as momentum fades. Similar to accumulation, there will be more sideways movement as bulls and bears vie for control. Volatility typically increases, corrections are deeper, and recovery is slower.

Smart money goes on a gradual selling spree throughout distribution, so as not to wipe out the order books. When there aren't enough bids to buy from retailers to absorb the supply, smart money ends up selling for less than they planned. *Smart money is always stealthy in their selling*, blending in as much as possible so they don't trigger any alarm bells for retailers.

Figure 80: Near the end of distribution, the uptrend is officially broken when the price makes a lower low. Selling volume is steadily rising, and is significantly higher than it was in previous weeks. Even though these signs may seem obvious in hindsight, they are easy to miss when complacency prevails.

Notice in Figure 80 that by the time the uptrend is broken, distribution is already 75% underway. This is one reason traders look for confirmation among several indicators. Among hundreds of indicators in TradingView, one in particular that is worth mentioning is the **relative strength index** (RSI): a momentum oscillator that compares the speed and change of a coin's price over the past 14 days. See Figure 81. Even though the price was increasing until mid-April, there was a bearish divergence between the price and the RSI that began three months earlier, in early January. This was an early indication that distribution was already underway.

Figure 81: The Relative Strength Index (RSI) can sometimes be an early indicator of a transition from markup to distribution.

Decline

The fourth and final stage of a market cycle is known as **decline**. It begins when bears take control, and the price plummets below previous support on high volume. Lower highs and lower lows begin to appear, and relief rallies are weak and unable to stand up to downward selling pressure. A clear downtrend is formed, with moving averages sloping down. Finally, short term moving averages will typically stay below longer term averages.

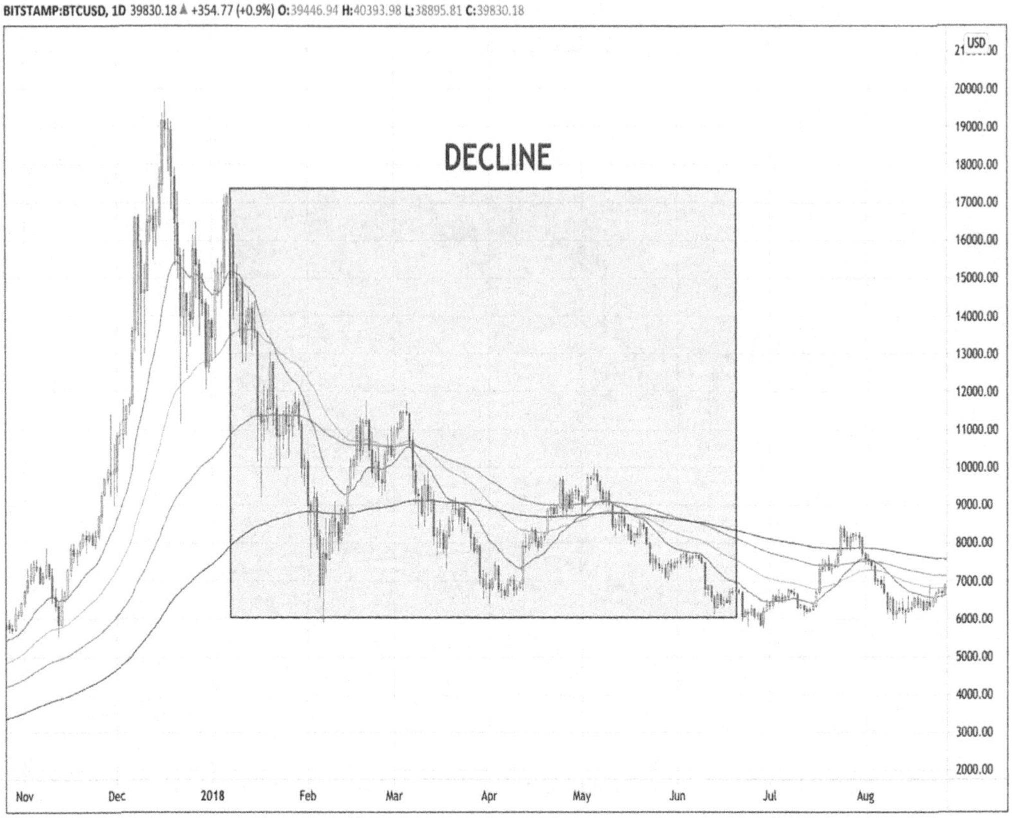

Figure 82: Decline, when the red line slips into the bottomless pit of FUD.

There may be a series of short relief rallies during a decline, where the price rebounds anywhere from 5-35%. This creates a **bouncing ball effect**: each time the price drops, some of its force dissipates as it hits the bottom, causing it to bounce back up with less momentum than before. The bouncing ball effect is what creates a series of lower highs and lower lows. With each pause in the downtrend, as long as the price doesn't rise above previous highs, you can assume that the trend will resume its course. *Declines are skeptical in the presence of rallies.* In other words, rallies are like tender saplings trampled underfoot by the reckless bear. This is the exact opposite of markup.

Trading with Cycles

All four stages are equally important to the conscious crypto investor. Accumulation is when the prices are at a premium, and is therefore the optimum time to buy. At the same time, there is higher risk involved, since you can never be 100% certain that a bottom has formed. In some bear markets, the price can consolidate for several months before suddenly dipping lower (Figure 83). More aggressive investors will set low bids they hope will be filled each time a significant dip occurs. They will strategically DCA over weeks and months, buying during each successive dip.

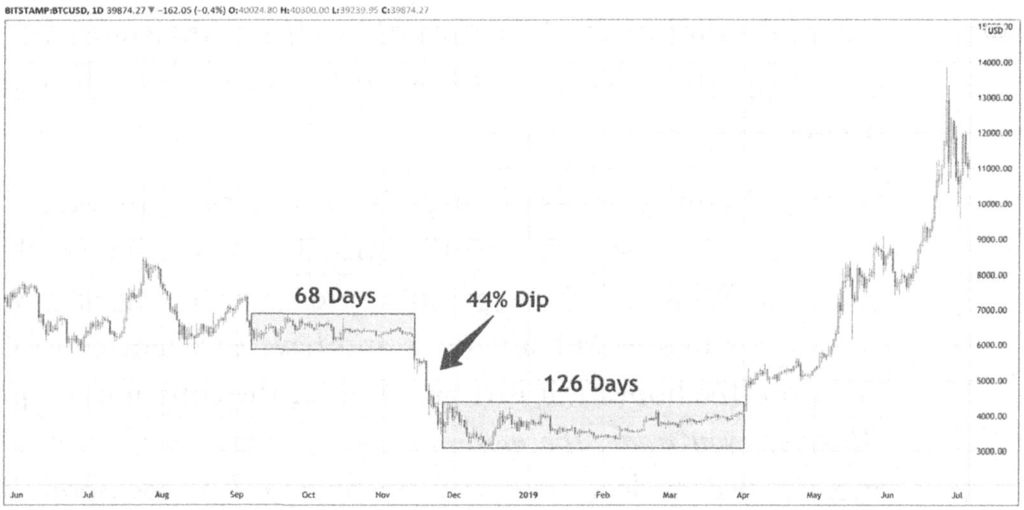

Figure 83: Notice how during the accumulation stage of the 2018-19 bear market there were 68 days of consolidation, followed by another 44% drop, and then a second 126 day consolidation period. Smart money would have placed bids in both zones, doubling down after the first dip.

Conservative investors with more to lose in the short term might wait until the early stages of markup, when they see a definitive breakout to the upside. After a long accumulation, a big green candle with significant volume on breakout means a higher likelihood of a trend reversal. Some traders will also wait a couple of days after the first close of a breakout for further confirmation.

During markup, more active swing traders may opt to take some profit off the table each time there is a big price surge. For example, they may sell their original investment when the price doubles. If they invest $1,000, they will sell their initial investment once the price rises to $2,000. This strategy helps to ease emotions, since the remaining capital is considered house money.

At this point, they may choose to sell a percentage–say 15%–each time there is a significant uptick in price. This way, by the time the uptrend is complete, they have sold everything and can patiently wait for accumulation to make its return. If by chance they don't sell everything at the top, they will have at least locked in most of their gains before decline.

Another more aggressive strategy is to "buy the dips and sell the rips." This means buying each time the price dips, and selling each time there is a surge. While this strategy can be profitable for day and swing traders who thrive in stressful situations and have excellent control over their emotions (be honest), it isn't practical for the conscious crypto investor. *The more you trade, the more mistakes you are likely to make.* Even the most skilled traders often end up buying dips too soon, and exiting trades too quickly. The potential reward isn't high enough to justify the extra time, stress, taxes and trading fees.

To summarize, accumulation and distribution are sideways trends, markup is an uptrend, and decline is a downtrend. Smart money will buy either during accumulation and/or during the transition into markup. They will sell all throughout distribution. During markup and decline–

typically at least half the duration of a market cycle–they will sit back and wait patiently, planning their next move. This is front-running the market in its broadest sense: getting in and slipping out before everyone else.

Sounds easy enough, right? It is, if you have discipline, patience, the ability to spot and follow the underlying trend, and a dash of intuition for good measure. Of course hindsight is 20/20, and since market cycles are never exactly the same, it's more challenging to spot transitions out of each stage as they are unfolding. Still, it is possible, if you pay close attention to changes in price and volume, and wait patiently for confirmation.

Trend Reversals

Trend trading involves timing exits and entries according to changes in the overall trend. Now that we have sufficient TA knowledge, we can explore **trend reversals** in more detail. There are both bullish and bearish trend reversals. A **bullish reversal** happens after a breakout above a downtrend line. A **bearish reversal** happens after a breakdown below an uptrend line. The ability to identify a trend reversal as it is unfolding is one of the most important skills that an investor can have. Fortunately, the past gives us all the data we need to master reversals. Let's look at a few examples to give you a sense for what to watch.

Notice the strong downtrend in Figure 84, which occurred during the bear market of 2014-15. The trendline is significant since it has been active for ten months, and there are multiple points of contact.

Figure 84: A bullish trend reversal during the accumulation stage of the 2015 bear market. Here there were 3 main signals that confirmed a breakout above the downtrend.

On June 16 the trendline was broken by a large green candle on high volume. These were the first two hints at a bullish reversal. Then on July 12 the price temporarily made a higher high, clearing the horizontal resistance line at around $300. Even if you had waited to buy in after the third reversal signal, you would have invested near the bottom of the bear market.

Now let's look for signs of bearish reversal during a strong uptrend in the bull market of 2017 (Figure 85). The parabolic rise from November 13 to December 16, where the price rose 260%, was the first warning sign. Big price movements in a short time period is a clue for smart money to be on high alert. The second sign was a bearish divergence between price and volume, with the volume steadily decreasing while the price continued to increase. The third warning sign was a spike in volume during the correction that ensued, which was higher than several previous weeks of green candles. This signaled that the bears were gaining control over the bulls.

The fourth sign of a bearish reversal was the magnitude of the correction. Not only did bitcoin drop 43% in six days, but it dropped beneath the low from the previous correction. The fifth and final sign occurred when the price wicked a significant distance below the uptrend line.

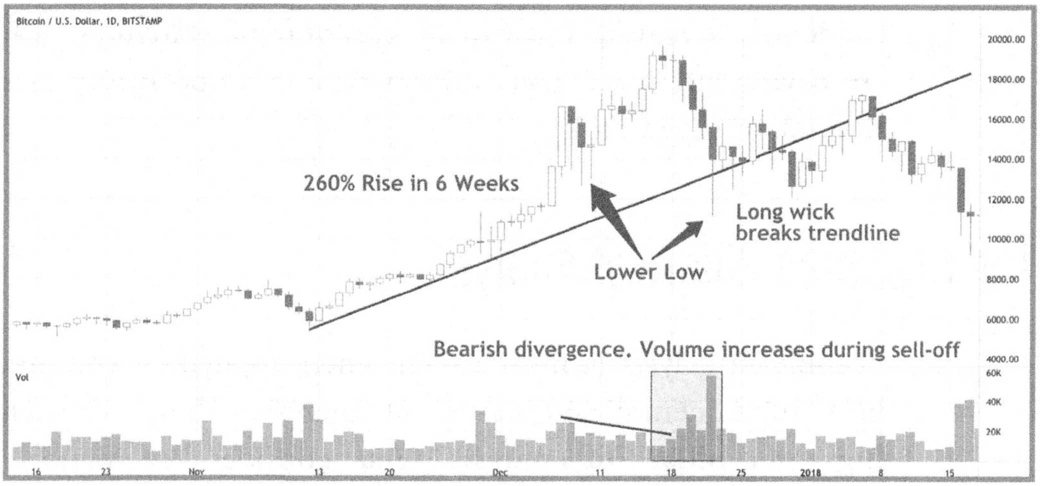

Figure 85: A bearish trend reversal after the distribution stage of the 2017 bull market. Here there were five signals that confirmed a breakdown below the uptrend.

Assuming you didn't catch the trend reversal until the fifth signal, you *still* would have had an opportunity to exit at around $16,000 during the next bounce, where the price temporarily recovered to over $17,000 on January 6 before making its final descent. Here we can see the bouncing ball effect play out in the form of a **dead cat bounce**: a bull trap after a blow off the top, where the price spikes back up temporarily after a sharp decline (the second hump in Figure 85). This gruesome analogy is given because it seems to be a knee-jerk response for so many bulls that never fails to play out, even after the most parabolic rises.

Unfortunately, residual euphoric delusion caused some bulls to ignore all the warning signs and rush in to take their positions. They paid for their impulsive entries during the long, cold winter (bear market) that followed in 2018-19.

While every trend will have different qualities, the two trend reversals above each reveal multiple different signals that trend traders are trained to identify. Regardless of which chart you are watching, as long as you are paying attention, you will have ample opportunity to respond to reversals as they unfold.

The Big Three Pillars of Analysis

Technical analysis is most useful when applied in combination with both fundamental and sentiment analysis. These are smart money's **big three pillars of analysis**. First, consider key metrics, use cases, developers and marketing, tokenomics, etc., to determine a potential candidate for investment. Next, track changes in social metrics, and/or follow crypto news and CT. Finally, observe price movements on a 1 year or longer timeframe. This will reveal personality profiles, which you can then measure against your risk tolerance to determine worthwhile investments.

You will likely notice that you tend to lean heavier toward one pillar over the others. It may be that it is naturally more appealing to you because it seems simpler and more practical to apply. Smart money will resist the tendency to become complacent with areas of strength, instead seeking to explore weaknesses that the average investor would avoid altogether.

Remember that each pillar has its own inherent limitations. If you rely on one pillar to the exclusion of the others, you are limiting your view. For example, if you find yourself constantly combing through news sources and social feeds to study market sentiment, you may have a good grasp about the current stage of the market cycle. But without looking at the fundamentals, you risk picking projects that have very little, if any long term potential. Similarly, without including TA as one of your allies, you are far less likely to be able to pinpoint when a systematic correction is approaching.

Finding the middle way between the three pillars of analysis means spreading out your focus. Avoid giving either too little, or too much attention to any one pillar.

Part 4
Secure Your Bank

Chapter 28

Private Browsing

While it may seem daunting at first, securing your cryptocurrency is really quite simple once you understand the basics. In this chapter we'll be reviewing the four most important features of security: private browsing, VPNs, standalone password managers, and two factor authentication.

The first order of business in the realm of crypto security is to download and install the Brave browser on your computer and mobile phone. Brave is a standard free web browser that looks almost exactly like Google Chrome, the world's top ranked browser. The difference is that Brave loads pages three times faster! Speed increases result from the elimination of ads and ad trackers, which greatly reduces the amount of content downloaded from websites. Every time you open a new tab on your browser, it will show you how many trackers and ads you've blocked, and how much bandwidth and time you've saved. This sets a new standard for privacy, since you keep the rights to your data.

Figure 86: Brave browser is the safest, most secure internet browser for cryptonauts.

Brave takes a radical departure from the traditional advertising model, which is characterized by involuntary data theft. Brave's advertising is both opt-in and privacy centric. You only see ads if you choose to, and you are incentivized to do so by receiving Basic Attention Tokens (BAT). BAT is Brave's ERC-20 token that is rewarded to users in return for their attention. Advertisers also benefit with better targeting and a higher return on investment, since viewers actually choose to see their ads.

There is also a Brave rewards program, where content creators and publishers can earn BAT tokens for creating meaningful content. This feature is already built into websites such as Twitter, The Washington Post, and The Guardian, to name a few. Brave is designed to reward the user, advertiser, and publisher. BAT has been around since 2017, and is traded on all major exchanges.

Virtual Private Networks

In addition to using Brave browser, consider adding a **virtual private network** (VPN) that uses extra encryption to protect both your data and your identity. A VPN enhances security by allowing you to set your IP address to anywhere in the world. This keeps you anonymous, protecting you from scammers, thieves, spies, and hackers. You can also bypass blocked websites and internet filters that may be censored by location. They are very affordable, and can be accessed from remote locations with ease. Two of the best VPNs available are ExpressVPN and NordVPN, both of which have desktop and mobile versions that run seamlessly in the background.

The only downside to using a VPN is that it can sometimes slow down your internet connection. If you only use a VPN when you are dealing with crypto, be aware that every time you turn it on, you may need to log back into your accounts, since your IP address has changed. Also be sure to use an accepted location, which will be specific to the exchange. For example, you wouldn't try to access binance.us from Switzerland. Finally, keep in mind that the language displayed on search engines, as well as the search results displayed, will default to the preferences of each country.

Password Managers

It is essential that you use a stand-alone password manager to store all of your passwords. Not only will it simplify the process of logging in to all crypto-related accounts, they have special features that significantly enhance your security. While many browsers such as Chrome, Firefox and Safari already have built-in password managers, they are limited in their functionality compared to standalone password managers. LastPass is highly recommended, for the following reasons:

1. It can be used across several different devices, and integrated with several different browsers. It is easy to import your stored passwords from existing browsers.

2. It can both suggest strong passwords and store them at the same time, allowing you to maintain hundreds of unique passwords. This way, if one of your accounts is hacked, the others won't be affected.

3. It uses the highest level of encryption. LastPass is transparent about how your data is protected, and has never suffered a breach.

4. Built-in sharing features on family plans allow you to share updated passwords with family members, who may need to access your accounts.

5. The LastPass vault is encrypted, and secured by a master password set by you. You can also add biometric authentication–face or touch ID–along with two factor authentication (covered in the next section) as additional security measures.

6. LastPass highlights passwords that are weak or used on other sites. You will also be notified if a password you are using is part of a leaked database, to prevent fraud.

7. You can store much more than just passwords: you can create secure notes containing your wallet PIN or public keys, or add attachments

such as driver's license and passport photos to be used for KYC verification on exchanges.

LastPass has both a mobile app and desktop browser extension, both of which run in the background once you've logged in with your master password. Each time you either create or update a username and/or password, a popup will appear that enables you to save the information in your vault for later use. Then whenever you revisit the site, your credentials will automatically appear in the input fields. You can even store credentials for multiple accounts on the same website. If you hover over the LastPass icon at the far right of the input field, you can quickly switch accounts. While it may seem like a few extra steps to get you set up, it will save you a great deal of time, energy, and frustration for potentially decades to come.

Use 2FA for Everything

Two factor authentication (2FA), is an added layer of security that checks to make sure you are who you say you are. With a steady rise in cyberattacks leading to significant data breaches over the past decade, it's clear that we need more than just a username and password to secure our internet identity.

We already use 2FA for pretty much everything these days. As an example, whenever you log in to your bank account, they will often ask for a PIN or ask you to answer security questions. If you use a new IP address, they may opt to use SMS or email to send you a one time access code to verify that it's you. These factors are all pieces of information that identify you in order to keep your accounts secure.

Google Authenticator, the original 2FA app, is a free mobile app that creates a time sensitive passcode, or **soft token**, that refreshes every 30 seconds. This type of 2FA token eliminates the chance of hacker interception, since it is both generated and displayed on the same device. *2FA tokens can, and should be generated for all crypto-related accounts that you use, without exception.* In fact, most exchanges will include this step as a critical part of setup in order to secure your account.

The free Authy app by Twilio is a major step up from Google Authenticator, since it enables you to sync your 2FA accounts across multiple devices, offers cloud-backups, has biometric recognition with face and touch ID, and has design features that allow you to link website icons to your codes for clearer visual recognition. Its website also has guides that give you step-by-step instructions for enabling 2FA for a variety of websites. Most importantly, it is possible to recover your account if you lose your phone. With Google Authenticator, unless you've exported all of your 2FA keys to another phone, if you lose your phone then you need to go through a lengthy process to get back into your accounts and reset 2FA.

How to Use Authy

Once you've downloaded Authy and set your username and password, login to your exchange on your desktop and navigate to Settings -> Security -> 2FA, and you will see a page that resembles Figure 87. Note: whenever a site prompts you to use Google Authenticator (most do), you can substitute the Authy app in its place.

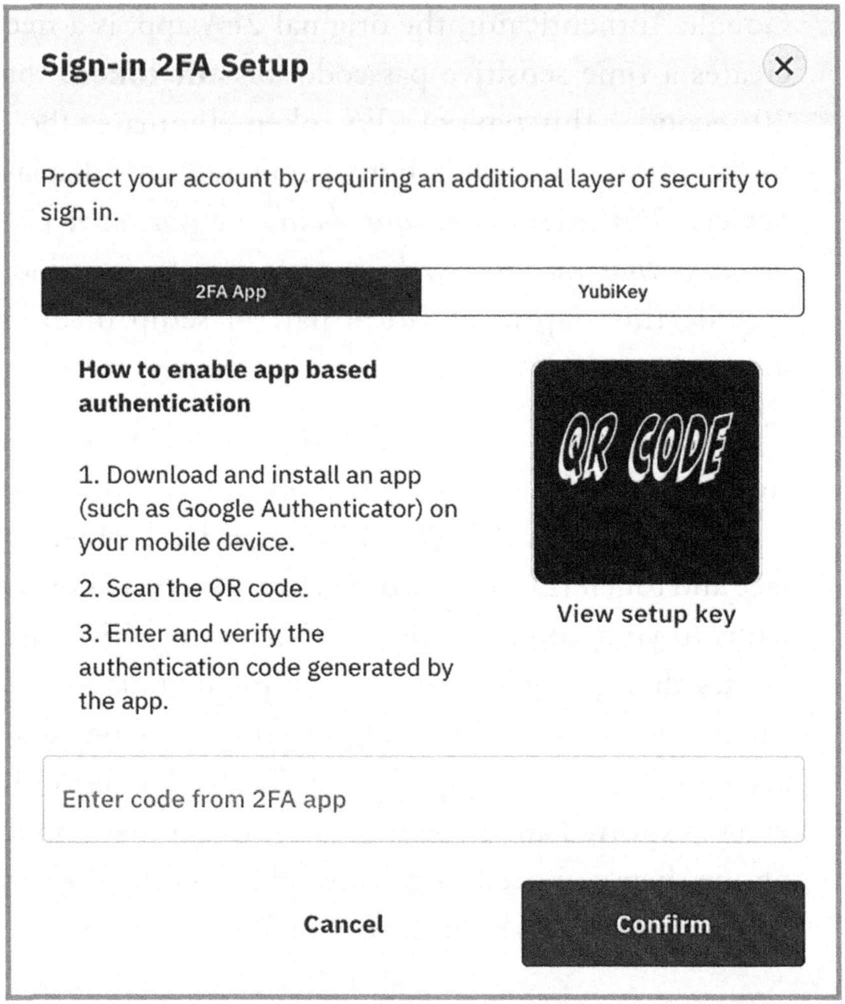

Figure 87: Setting up 2FA on Kraken exchange. All trading platforms have similar procedures for 2FA setup. Note that the QR code in this image has been covered.

For each exchange, it is possible to create separate 2FA tokens for sign-in, trading, and funding. Choose sign-in to start, then open Authy and tap the + symbol in the bottom right. You will then use your phone's camera to scan the QR code shown on the exchange. The connection is confirmed after you enter the corresponding six digit token that pops up on your device.

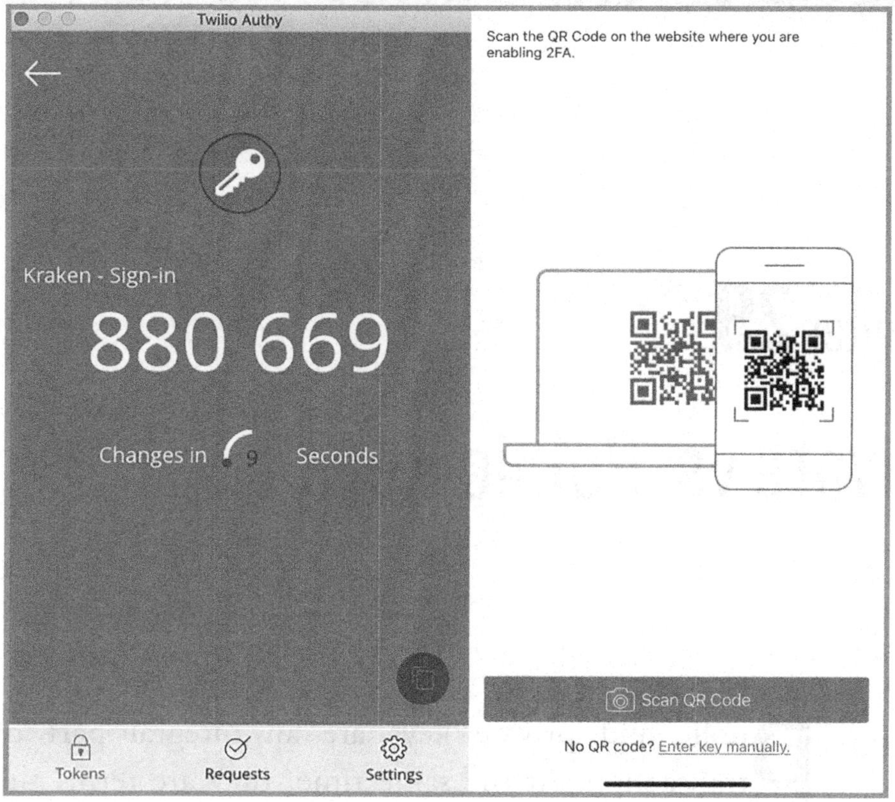

Figure 88: Open your Authy app and tap the + to add a new account. Then either choose to scan a QR code or copy and paste the code manually.

Note that the 2FA token for each site will refresh every 30 seconds, and that you must enter it within that window in order for it to be valid. If there are less than 5 seconds left, it's better to simply wait until the next token appears to save you the frustration of having to enter it twice.

Once you've established a 2FA link, each time you attempt to login, you will be prompted for the soft token from your Authy app. My personal preference is to only create 2FA tokens for sign-in and funding, and leave it off for trading, since I don't think it's necessary or helpful when I'm trying to execute market orders. No matter what you choose, always use 2FA for sign-in!

Chapter 29

Private vs. Public Keys

Public and private keys are an integral part of every crypto transaction. At the same time, they are terms that can be easily misunderstood. A **private key** is an encrypted alphanumeric code: a mathematical string of letters and numbers that grants access to holdings for a specific cryptocurrency. Private keys prove ownership, in the same way that a password gives access to an account. They allow you to make transactions and spend your holdings. If anyone else has access to your private keys, they can do the same. As cryptocurrency advocate and author Andreas Antonopoulos is fond of saying, "Not your keys, not your crypto!"

When a private key is created, it will automatically generate a related **public key**. Private and public keys are linked mathematically, and are referred to as **key pairs**. Public keys are pseudonymous, since they are linked to you but don't reveal information about your personal identity. Your **public address**, which you will use to receive crypto, is a hashed version of your public key. A public address is often compared to an email address, but it is more like a transparent PO box. Anyone who has your public address can send you crypto. They can also use blockchain explorers to see how much is moved into or out of your public address. But without your private key, they can't open the box/remove your funds.

Since every cryptocurrency blockchain has its own type of address, it is necessary to have a key pair for every coin or token that you own. Fortunately today we have crypto wallets, which allow you to access all of your key pairs in one convenient place.

Chapter 30

Crypto Wallets

A physical wallet is a universally recognized symbol for money storage, one which has been around for as long as paper money has been trading hands. In light of this, early adopters of cryptocurrency decided to confuse the hell out of everyone by applying the concept to digital money storage. It was a convenient, but poor choice, since crypto wallets don't actually store crypto–they simply hold the private keys which you can use to access your holdings. If you "lose" the wallet, whether it be a secure hardware device or your phone where an app is stored, you don't actually lose your crypto.

For this reason, a more sensible term for a crypto wallet would be a **keychain**, since it holds your key pairs. You can make copies of your keys and transfer them to as many keychains as you'd like. We will use the terms keychain and wallet interchangeably, since in terms of cryptocurrency they both refer to the same thing.

When a keychain is created, it generates a **root seed**, which is a very long number from which countless key pairs can be generated. Since the root seed would be very difficult to remember, keychains translate them into a **seed phrase**: a series of 12 or 24 words that represent your root seed in plain english. Seed phrases are a gateway to your key pairs, and therefore your crypto holdings. A seed phrase is sometimes also referred to as a recovery phrase or a secret passphrase.

The key pairs that you access with your seed phrase can be stored on multiple keychains. If the keychain is lost, then you can easily import your keys again using your seed phrase. *This is why you must write down and safely store your seed phrase to secure your crypto.* Ideally you will store a minimum of two physical copies of your seed phrase in two distinct locations. One example would be a fireproof lockbox that is bolted to a cement floor. *Do not post your seed phrase on the internet, or give it out to friends.* Crypto exchanges or other related providers should **NEVER** ask you for your private keys, for any reason. If they do, it's 100% a scam.

Every cryptocurrency will have its own private and public address that is determined by its unique hashing algorithm. Each set of keys therefore speaks a unique language. For this reason, you never want to send ADA coins to a BTC address, because their blockchains don't understand one another. Blockchains also don't have lists of valid addresses, which means it is possible to send coins to addresses that don't exist. If you send coins to addresses on the wrong blockchain, or you make a typo when sending to an address, those coins are burned, or lost forever.

Sounds like a terrifying prospect, one that can make you feel uneasy about moving your crypto off the exchange and onto a keychain. Fortunately, you can completely sidestep the issues mentioned above by taking a handful of simple precautions. We will explore these in greater detail later in this chapter.

Since key pairs and keychains are terribly confusing to the average crypto user, most prefer to keep their crypto on exchanges. But the small convenience this provides doesn't justify the risk. Wallets that are connected to exchanges are custodial, which means if the exchange is hacked, its servers go down, or your account is frozen for any reason, your funds will be at risk. To eliminate this possibility, use a **non-custodial wallet**, where your private keys are solely under your custody.

Conscious crypto investors understand the great value in learning to store crypto safely and securely for decades to come. Once you have done a few simple transfers, you will find it as simple as moving money into and out of your bank–and in a fraction of the time!

Software Wallets

When moving your coins off an exchange, there are two two main categories of crypto wallets to choose from: software and hardware. **Software wallets** are also referred to as hot wallets or mobile wallets, since your access to them is online. **Hardware wallets** are often referred to as cold wallets: the keys are accessed offline, on a secure device that resembles a USB drive, but with far superior security features. There are pros and cons to both software and hardware wallets.

Software wallets are easy to set up, and you can transfer various cryptocurrencies with relative ease. If you lose your phone, you can easily download the application again and initiate your wallet from an existing seed phrase. They are user friendly, and most have extra features that include buying crypto directly within the wallet, swapping tokens on specific networks, staking, and NFT storage.

Trust wallet and Exodus wallet are very popular software wallets with hundreds of thousands of users. Both are free and easy to use, and have excellent customer support. Trust wallet is the clear winner when it comes to support for the largest number of cryptocurrencies. At the same time, Exodus wallet has a sleek desktop version, while trust wallet currently does not. The ability to access your wallet from your computer allows for a foolproof way to copy and paste addresses so that you can move your crypto efficiently. This is especially helpful when you want to sell some of your positions fast!

The downside to software wallets is that your private keys are stored within the app itself. This means that if your phone is hacked or infected with malware, there is a chance that your keys can be stolen, and your wallet wiped clean. While this isn't a risk with very high probability, it is definitely something to consider. This is one reason why conscious investors will spread their holdings across multiple software wallets, and/or opt to use a hardware wallet. We will explore the key features of hardware wallets in the next section. First, let's do a walk through for Trust wallet, since the setup is nearly the same for most other mobile wallets.

Step 1: Download and Install the Trust Wallet App

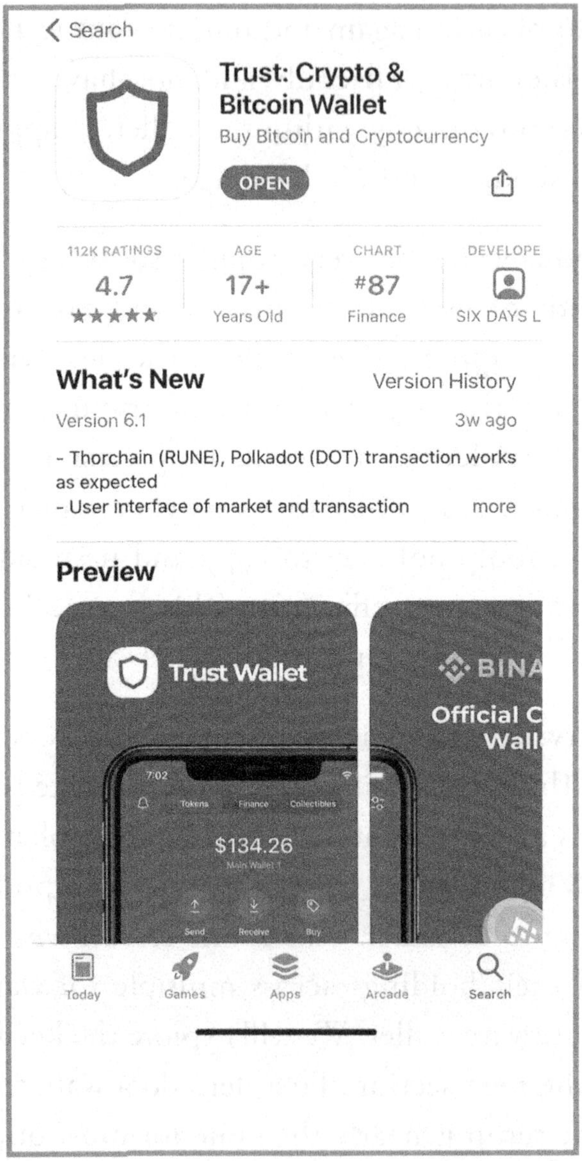

Figure 89: Trust wallet in the iphone app store.

Step 2: Choose to Create a New Wallet

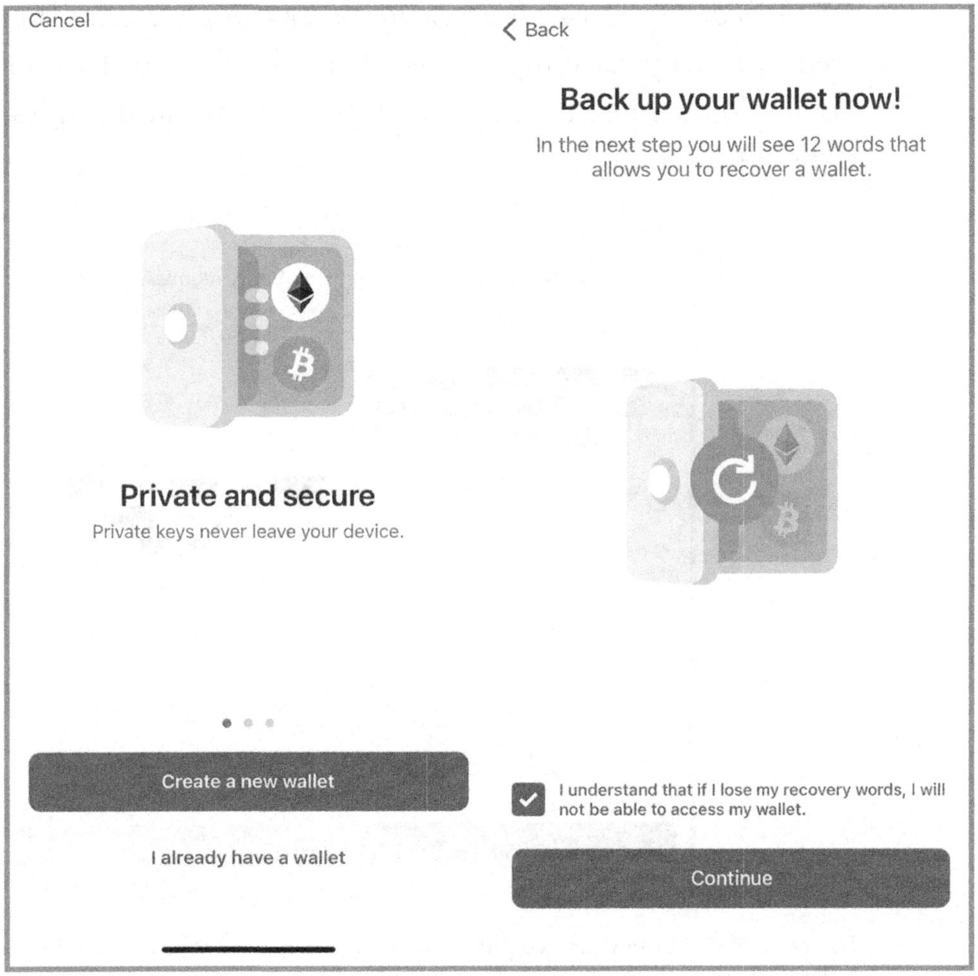

Figure 90: After choosing the option to create a new wallet, you will be encouraged to back up your seed phrase.

Choose the option to create a new wallet, which will generate a unique seed phrase for you to write down. If by chance you lose or break your phone in the future, and need to reinstall the app, you will choose "I already have a wallet."

Step 3: Write Down Your Seed Phrase

Write down your seed phrase in the order it is presented. Then verify your seed phrase by selecting the words in the same order you wrote them down. Remember to create at least two physical copies, to store in different places.

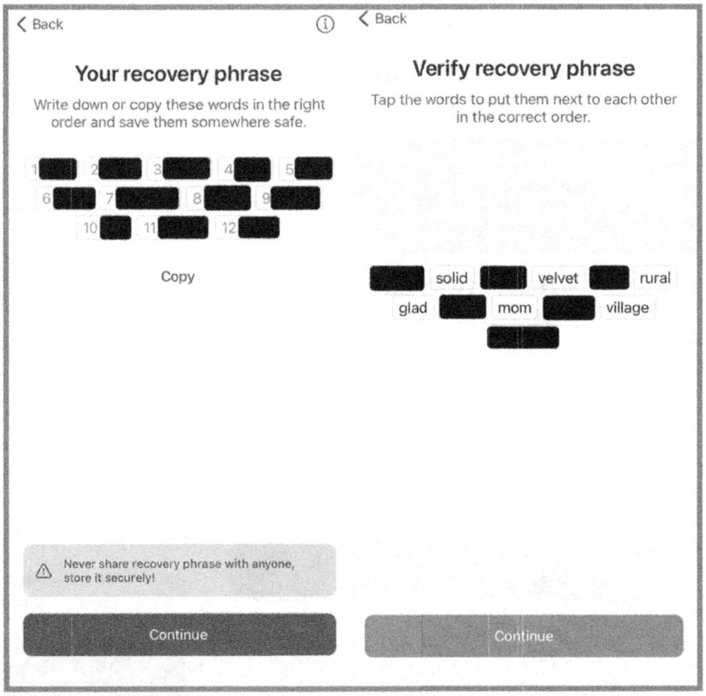

Figure 91: Write down your seed phrase twice, and then verify it by tapping the words in the correct order. Note that the phrase above has been concealed to prevent anyone from using it to store their crypto!

Step 4: Generate Relevant Key Pairs

Once you've set up your wallet, you'll notice that there are already a handful of cryptocurrencies that are automatically featured (Figure 92).

Tap the list icon on the upper right of the dashboard to manage which to display. You can search by either full name or ticker, and slide the button to the left to activate the key pair. You will want to activate all the cryptocurrencies that you plan to store in your wallet, so that you can retrieve their addresses.

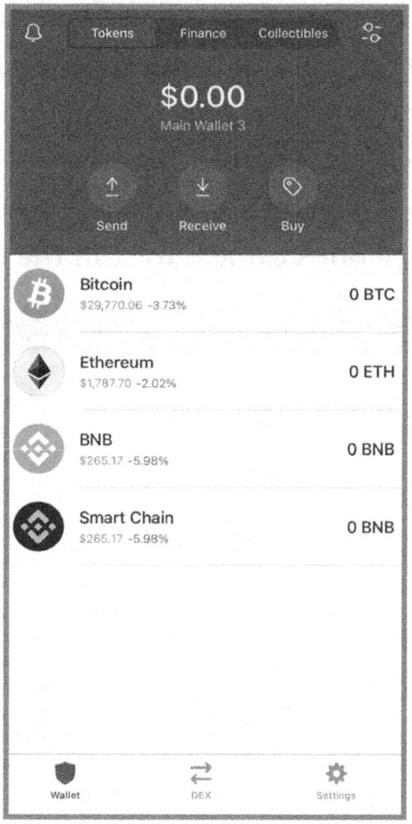

Figure 92: All mobile wallets have a handful of key pairs automatically activated at startup. You can easily toggle addresses for various cryptocurrencies on and off by tapping the list icon in the top right of the wallet tab.

Note that even if you were to deselect a cryptocurrency that you own, the balance would still be stored on your keychain–it just wouldn't be visible from your dashboard. This applies to all other wallets as well. An example of this would be MetaMask wallet, which is strictly for ERC-20 tokens. MetaMask is the preferred way for some to conveniently connect to DEXes such as Uniswap and Sushiswap. For every transaction you will be sending and receiving from the same ETH address. Your balances won't appear until you search for and then import their respective tokens.

Once you've activated a token in your dashboard, you can access its public keys by tapping it and then selecting "Receive" (Figure 93). Both the QR code and the alphanumeric series below it are revealing the same public address. Which one you use will be a matter of preference and convenience. For example, if you are sending crypto from your mobile wallet to an exchange account on your computer, you may find it easier to use your phone's camera to scan the associated QR code.

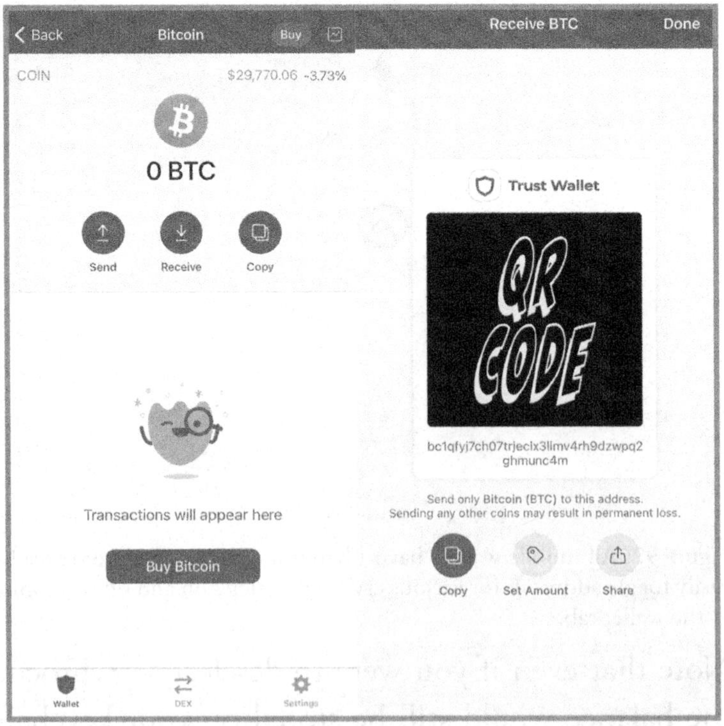

Figure 93: After tapping the "Receive" icon for BTC, you will see your public address displayed as both a QR code and an alphanumeric sequence. They are both the same.

There will also be times when you won't be able to scan the QR code. An example of this would be when you are sending crypto from the exchange to your phone. In this case it is easier to hit the copy button that appears at the bottom left of the address, and paste it into your wallet.

There are other features that you may find useful within Trust wallet, such as a built-in DEX that allows you to easily swap certain tokens (Figure 94). For example, you could swap ETH for AAVE (both ERC-20), or BNB for RUNE (Both BEP-2). Make sure that you always have some ETH and BNB in your wallet to cover gas fees.

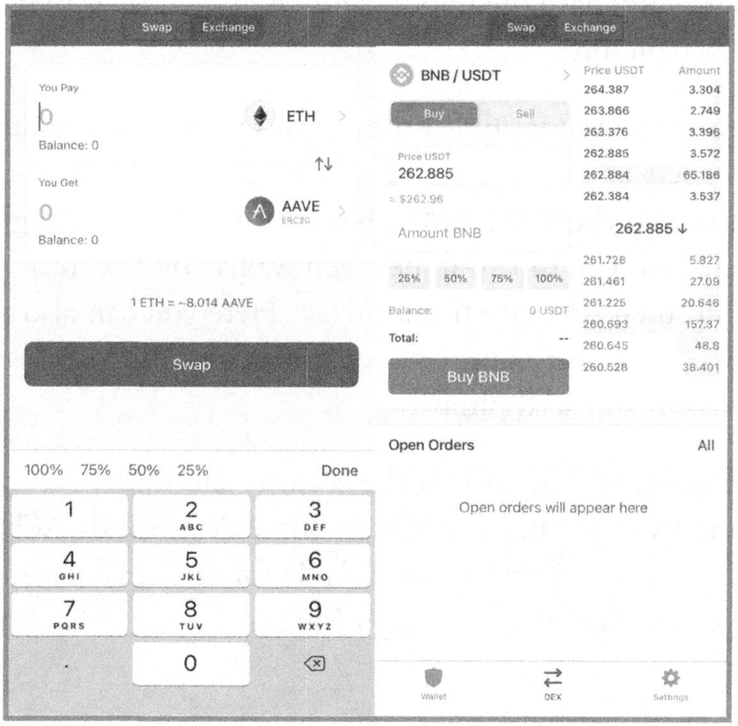

Figure 94: Trust wallet's swap and DEX features. You pay a premium in exchange for the convenience of not needing to move your crypto around.

There is also an exchange tab where you can use stablecoins to purchase certain ERC-20 or BEP-2 tokens. The available tokens to trade will continue to expand as the technology evolves. For instance, Thorchain (RUNE) is developing a cross-chain DEX that will allow users to swap coins across multiple blockchains. It may eventually be integrated into Trust wallet, among others. Think how much easier it will be to not need to send your crypto on and off exchanges to swap tokens!

You might be wondering why exchanges are necessary if you can buy and store crypto straight from your wallet. While these features are useful for smaller purchases, at present the fees are much higher, and the functionality is still very limited. This will continue to improve with time, but exchanges currently have the upper hand because of their higher liquidity and lower fee structure.

Another important feature of Trust wallet is the ability to manage multiple wallets at once. If you go to Settings -> Wallets -> +, it will walk you through the process of either creating or importing another wallet. Once created, you can swap between wallets by returning to Settings -> Wallets and then selecting from the list. Here you can also rename each wallet or access recovery phrases. Remember that this isn't a substitute for writing down your seed phrase!

Finally, be sure to visit the security tab under settings, and activate both the Passcode and FaceID as your lock methods. While FaceID will be the first resort, if that is unavailable for any reason you will be able to open your wallet using a six digit PIN.

Hardware Wallets

Hardware wallets are secure devices that give access to private keys. They are considered cold storage since they don't need to be connected to the internet to be accessed. Again, your crypto isn't actually stored on the wallet itself, so if you lose it, it's no big deal *as long as you have your seed phrase.*

Trezor and ledger are two of the most popular hardware wallets that will allow you to store your crypto. In both cases, you must switch on your device, connecting to your computer either via bluetooth or with a USB cable. Earlier models use a small screen display that doesn't require you to connect to your computer. In either case, you press two small buttons designed for a 6 year-old child's fingers (seriously guys).

These wallets also allow you to buy directly using a third party, again with higher fees. Keep in mind that you must create a PIN that you enter each time you want to access the wallet, which is a necessary security feature that takes the place of biometric scans on mobile wallets. The option to connect your wallet to a desktop app makes it easier to see your portfolio, and speeds up the process of transferring tokens.

Pros and Cons

Hardware wallets tend to feel more clunky for the tech savvy. More steps to access your funds means it takes longer to transfer to and from exchanges. While some crypto users refuse to even touch software wallets for security reasons, a hybrid approach seems to make the most sense. Software wallets are useful for quickly and easily accessing crypto for more immediate trading. Hardware wallets, on the other hand, are best used for the HODL fund portion of your portfolio that you don't plan to move around anytime soon.

Ultimately you must decide how many and which type of wallets to use to store your funds. Some investors have five or more wallets, while others only have one or two. Balancing simplicity and security is the key.

9 Helpful Precautions

Now that you are familiar with the basics of crypto wallets, let's explore some precautions that are both helpful and necessary to ensure that your holdings stay safe and secure at all times.

Precaution #1: Never move crypto to or from a wallet when you are tired, in a frenzy, high on anything, or low on alcohol. You are your own bank, and greater freedom comes at the cost of greater responsibility. Always be alert and aware when managing crypto. Simple mistakes are irreversible, and can cost you a fortune.

Precaution #2: Make absolutely certain that you are sending your crypto to the same network, and use a memo when required (Figure 95). Bitcoin goes to Bitcoin, Ethereum goes to Ethereum, etc. While these examples are relatively straightforward, there are some cryptos, such as RUNE, which have tokens that were issued using different network standards. RUNE has an ERC-20 version, a BEP-2 version, and a native version. How will you know the difference? This information will appear on the exchange where you purchased the coin, and can also be seen in the information section when you go to make a withdrawal. In fact, exchanges will clearly list the steps and precautions you should take when sending each coin. Be sure to read them!

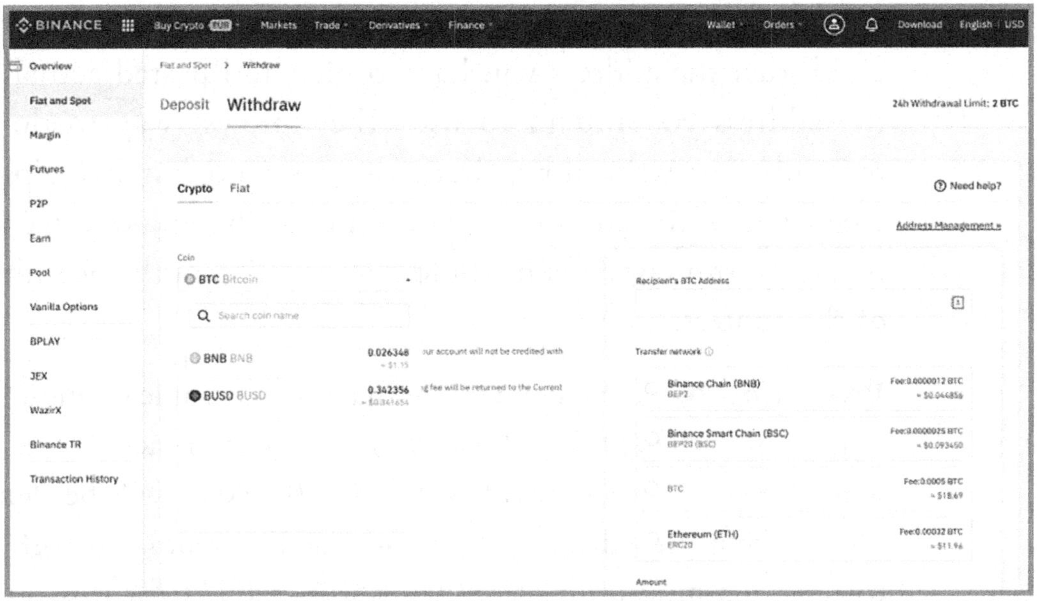

Figure 95: Binance's withdrawal page, which you can find under "Wallet." Choose the type of coin you want to withdraw, then paste your external wallet's public address. Finally, choose the matching network, and enter a memo if required.

When it comes to memos, the exchange or wallet will tell you whether one is required. If there is a memo containing an alphanumeric series in any of the withdrawal fields, don't forget to add that as well.

Precaution #3: Make sure that you are crystal clear about the To: and From: addresses. For example, if you are wanting to send crypto from your Kraken exchange to your Trust wallet, then you know that you will need to paste the "Receive" address in your Trust wallet into the "Withdraw to" address in your Kraken account.

Precaution #4: Always use either a QR code, or select the copy icon to transfer your public address. This will eliminate the chances of making a typo that you will most definitely regret later.

Precaution #5: For your own peace of mind, it is helpful to double check that the address you have copied and pasted actually matches the address you intend to use. The simplest way to do this is to remember or write down both the first and last five alphanumeric characters of the address. Chances are, in the unlikely event that you made an error copying, it would appear either at the beginning or end of the address.

Precaution #6: Always make sure that you have at least enough ETH on both your exchange(s) and wallet(s) to cover transaction fees, otherwise your transfer requests involving ERC-20 tokens will be denied. Since most tokens are currently built on the Ethereum network, it is wise to reserve a minimum of $200 - $500 of ETH for this purpose. The same rules apply for BNB, since it is used as gas for transactions on the Binance Smart Chain.

Precaution #7: Don't freak out when transactions take longer than expected. While some blockchains, such as Solana (SOL) can process thousands of **transactions per second** (tps), others such as Bitcoin are slower. Understand that each has its own consensus mechanisms and hash rates, with some requiring more network confirmations than others. Like bottlenecks that occur at exit ramps on the highway, there can also be delays in transaction processing due to congestion. Sometimes a transaction that should have appeared in ten minutes can take 24 hours or more (rare but can happen).

Every reputable exchange and wallet will be fully transparent about the status of a transaction. The status will change from pending when the request is made, to initiating, and then processed once the required number of confirmations is made. You can also track transactions by pasting the transaction hash from your order history into a matching blockchain explorer. In the unlikely event that a transaction isn't processed, whether it has to do with not paying enough fees or a network error, the crypto will simply stay in your wallet.

Precaution #8: If you feel uneasy about sending large amounts of crypto at one time, try sending a small amount as a test run. You will end up paying more in transaction fees, but it will take the stress out of the equation. Litecoin (LTC) would be a prime candidate for learning to send/receive crypto, since it has very low transaction fees (a few pennies), and block times of about 2 minutes. You could theoretically send $1 of LTC from your exchange to your wallet and back again in a few minutes. This will help you to practice and gain the necessary confidence to send larger amounts later.

Precaution #9: **Whitelist** public addresses that you use frequently on your exchange. Whitelisting involves letting your exchange(s) know which public addresses are safe to receive your crypto. Once saved, you can then select from a dropdown menu in the "Withdraw to" field. This saves the hassle of constantly having to copy/paste an address, and then verify its validity through email. Make sure to put the name of the cryptocurrency in the title of the whitelisted address, and double check the first and last five characters of the address against your wallet address before each use, just to be sure that nothing has changed.

Even though sending and receiving crypto using public addresses can seem confusing now, take heart! There are inter-blockchain protocols being developed, like the FIO protocol from the Foundation of Interwallet Operability, which will give wallets human-readable addresses to make sending crypto as easy as sending an email.

Manually Encrypting Your Seed Phrase

The problem with seed phrases is that while the chances of being guessed are slim to none, if anyone with even a basic understanding of how crypto works finds them, they can easily import your keys from anywhere in the world and sink your fiat lifeboat. How to avoid this?

The general consensus is that you should keep multiple copies of your seed phrase, storing at least one copy inside a fireproof lockbox in your house, and another in a safety deposit box at the bank. It isn't recommended that you store them on any cloud platform, since these aren't very secure.

You should definitely have several copies of your seed phrase stored in different places that are both safe and secure. However, the problem with these methods by themselves is that they are tedious, and not completely foolproof. Despite certain laws that are supposed to protect the contents of safety deposit boxes, there have been numerous instances of illegal seizure in recent years.[1] This doesn't mean you shouldn't use them, but there is a different way to hide your keys that will help you to rest easier at night.

The manual encryption method explained in this section adds an additional layer of security to your seed phrase that will stump any potential evesdroppers. As a disclaimer, the seed phrase used in this example is as fake as the fiat figure on your next stimulus check. In other words, you should never use it for any wallet! Got it? Good.

Let's say that you are given the following seed phrase:

1. Loan
2. Farfetched
3. Umbrella
4. Give
5. Whoever
6. Body
7. Keys
8. Not
9. Dead
10. My
11. Over
12. Steal

Now what would happen if you hypothetically swapped the 6th and 12th words in the phrase, as well as the 9th and 11th? Your seed phrase would now look like this:

1. Loan
2. Farfetched
3. Umbrella
4. Give
5. Whoever

6. Steal
7. Keys
8. Not
9. Over
10. My
11. Dead
12. Body

Now you have all the words from your seed phrase, with four of them out of order. Great, now what? Well, if you left it like this, chances are you'd quickly forget the actual order of your seed phrase! To help you remember, all you need to do is take it a step further by coming up with a **decipher code**.

The decipher code is an alphanumeric series that will help you to remember how you jumbled up the order of your seed phrase. In the above example, you might create something like this: 612A761ai927cwxB911

In the above example, you can see 126 at the beginning of the code, and 13 at the end. If you know that the code is showing specific numbers that have been swapped, and that each swap sequence either begins or ends with a capital letter, then you can easily remember the original order. The rest of the code serves as a distraction for eavesdroppers. If you don't trust your memory that well, you could greatly simplify the code and still have much better protection than if you did nothing at all. As an example, 612911 is still meaningless to outsiders.

Never Share Your Keys?

Die hard crypto enthusiasts will tell you ad nauseum to never share your private keys with anyone. Let's take a moment to amend that statement, for the sake of your loved ones. *Do not share your private keys with anyone* ***except*** *for those who will receive your inheritance.* For most people, this will likely be our children and/or spouse. There are an overwhelming number of stories of investors big and small who made unexpected trips to the grave without sharing their private keys. Their crypto is now somberly floating around somewhere on the blockchain for the rest of eternity, with no-one to call its owner. Remember that without access to your keys, no power on this good green Earth can help recover your funds. **Please share your keys with the right people**.

Chapter 31

Passive Income for HODLers

Many new investors are stunned when they hear about the possibility of leveraging their crypto to earn passive income. They are used to receiving a measly 0.04% APY on their savings, not realizing that there are a few simple methods to earn exponentially more. As an example, if you have spare cash lying around, and you don't want to deal with the volatility of the cryptosphere, you can easily convert it to a stablecoin and loan it out for an average 12.5x higher APY than you would receive from your bank.

Part of the appeal of crypto lending platforms is that you can choose the interest rate according to your preferred **lockup period**: the amount of time your crypto will be locked into smart contracts, and therefore unavailable for trading. If you choose a flexible lending term, you will receive a lower APY for the right to withdraw your funds at any time.

Crypto Lending and Yield Farming

How is it possible to earn so much on your crypto? There is a very high demand for loans in crypto, most of which comes from leverage traders. Some platforms also lend to crypto projects looking for early funding, which means that each time you lend, you are contributing to the evolution of the cryptosphere.

Those who are already invested in crypto and plan to HODL can lend out their coins for impressive interest rates. Imagine getting paid extra just to hold bitcoin, ethereum, or dozens of other altcoins that are already earning sizable returns over the long term! Visit defirate.com/lend/ to compare lending rates across a dozen centralized and decentralized crypto lending platforms. They offer rates for both stablecoins and over a hundred other altcoins, which are updated hourly. Rates change according to the supply relative to demand for each cryptocurrency.

The differences between centralized and decentralized lending platforms are similar to those of exchanges. Centralized lending platforms are basically custodial wallets where you deposit your funds. Examples are Celsius, Blockfi, and Nexo. You hand over your private keys in return for the security of using an insured and regulated platform. It is relatively easy to set up an account: download the app, perform KYC, and deposit coins into your wallet. In most cases you are paid out either daily or weekly, either in the cryptocurrency which you deposited, or in the native token issued by the platform. Nexo and Celsius, for instance, offer higher interest rates for loyalty members who choose to receive NEXO or CEL tokens, respectively.

Decentralized lending platforms such as Aave, Compound, Anchor Protocol, and CREAM Finance are open-source DeFi applications that also allow users to earn interest on deposits and/or borrow by putting up collateral. They each have billions of dollars in TVL, don't require KYC checks, and don't limit who can use them with geographic restrictions. Also, they enable you to deposit funds while retaining custody of your keys. Interest is compounded each time a block is mined, which for ERC-20 tokens such as Aave and Compound is every fifteen seconds.

In the same way that anyone can loan out their crypto by locking it into a lending protocol, it is also possible to lock up crypto as collateral in return for a loan. While this is not recommended for the purposes of creating leverage, it can be useful for other loans if you don't want to go through the traditional process of proving that you are creditworthy.

Taking out a loan against your crypto during a bull market is a smart money move, since the gains you receive can eventually pay off most of the loan. You start by creating a **Collateralized Debt Position** (CDP), and after locking up collateral in smart contracts, can withdraw a percentage in the form of a stablecoin loan. For most crypto loans, you will need to lock up at least twice as much crypto as you borrow.

Let's say that you owned $1,000 worth of LUNA, and the price of LUNA was currently $20. Using Anchor Protocol, you can bond your 50 LUNA by minting a liquid staking derivative called bLUNA. 1 bLUNA = 1 LUNA. After locking up your bLUNA as collateral, you will be able to withdraw up to 50%, or $500 of UST, LUNA's decentralized stablecoin.

The amount that you withdraw as a percentage of your total collateral is referred to as your **Loan to Value** (LTV) ratio. This ratio fluctuates along with the price of LUNA. If the price of LUNA goes up, then your LTV drops, and vice versa. As long as your LTV is 50% or less, your collateral is safe. As soon as it rises above 50%, your collateral is liquidated (max LTV subject to change). Note that you would still keep your initial loan.

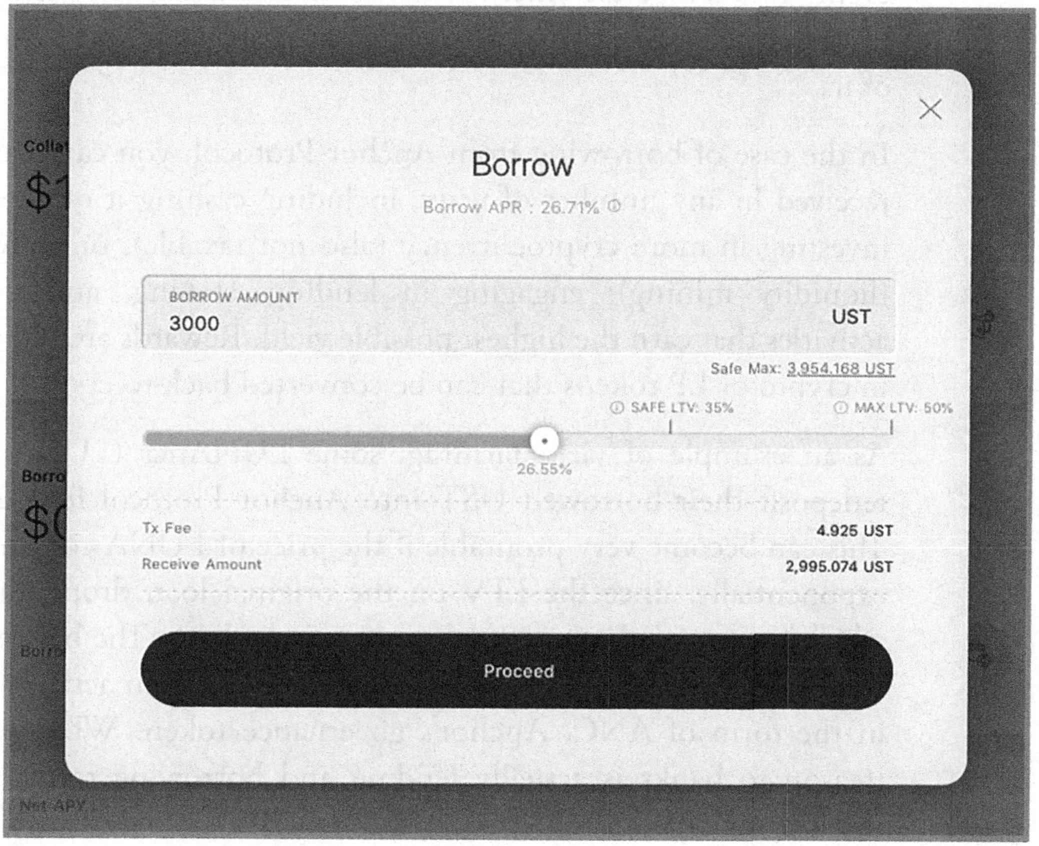

Figure 96: Borrowing using Anchor Protocol. Notice the value given for "Safe LTV" is 35% or less.

To prevent liquidation of your collateral, it is safer to choose a lower LTV. A conservative LTV would be 20% or less. This way, even if the price of LUNA suddenly dropped 50% to $10, your LTV would only rise to 40%, which is still below the liquidation threshold. Of course, you can also add collateral in the form of bLUNA to your debt position at any time, effectively lowering your LTV.

You might be wondering why anyone would want to put up exponentially more collateral than the loan they receive. There are several reasons. For starters, receiving a loan for depositing your crypto is not a taxable event. This can be extremely beneficial in a bull market, when you don't intend to sell your crypto, and are looking for clever ways to make use of it.

In the case of borrowing from Anchor Protocol, you can use the UST received in any number of ways, including cashing it out to pay bills, investing in more cryptocurrency (also not taxable), or **yield farming** (liquidity mining): engaging in lending, staking, and/or liquidity activities that earn the highest possible yield. Rewards are often paid out in crypto or LP tokens that can be converted back to crypto.

As an example of yield farming, some LUNAtics (LUNA fans) will redeposit their borrowed UST into Anchor Protocol for ~20% APY. This can become very profitable if the price of LUNA continues to rise exponentially, since the LTV on the original loan drops significantly, which means it takes much less to repay the loan. The best part is that instead of paying an APR on existing loans, you can actually earn APY in the form of ANC, Anchor's governance token. What sounds like treason to banks is actually lending and borrowing terms that fairly reward participants.

At the risk of sounding like Uncle Sam, there are a few downsides to decentralized lending platforms that you'll want to keep in mind before depositing your funds. They are not yet insured or regulated, and their rates are typically lower than their centralized counterparts. They also won't be able to help you recover funds that are lost for any reason. Examples of this would be a hack or a glitch in smart contract operations. Always DYOR, and make sure the platform is legit before depositing your crypto into their smart contracts.

Staking

Staking is an aspect of yield farming that involves locking crypto onto a network in return for rewards. Staking is possible with **Proof of Stake** (PoS) cryptocurrencies, which use a consensus mechanism where **validators** verify and secure transactions on a blockchain network. PoS differs from PoW in that validators with a stake in the network replace the need for miners. Should validators try to steal by corrupting the blockchain, they face severe slashing penalties.

The main benefit of PoS is there is no need for specialized computer hardware to solve complex cryptographic puzzles, which means less energy consumption and more transactions per second. The main pitfall of PoS is that it isn't as decentralized or secure as PoW, since validators with a larger stake in the network have more control. A larger stake also gives validators more sway when it comes to the governance of a project.

While most cryptos require you to put down a hefty chunk to become a validator, anyone can become a **delegator** by locking coins into a **stake pool**: an address that is run by a validator. Validators then share their rewards with delegators according to how much they have staked in the pool, minus a delegator fee.

Staking has become so popular with crypto investors that most wallets and exchanges allow you to directly stake some or all of your holdings for select cryptocurrencies. Be aware that there may be lockup periods for some networks that will penalize you for removing funds early.

Stakingrewards.com is a one stop site for all things staking. It conveniently ranks coins by staked value, providing average reward percentages, as well as the percentage of coins that are being staked. Solana, for example, has 76% of its circulating supply staked on the network at the time of writing. A high percentage staked is a strong vote of support by the community, and can serve as a buffer against larger price dips during systematic corrections.

On each project page there is a nifty calculator that will determine how much you can earn monthly and annually based on the current value of the coins you plan to stake. You can also compare reward rates for delegating with different validator nodes, read a FAQ that explains how to stake on the network, and learn about potential risks or lockup periods.

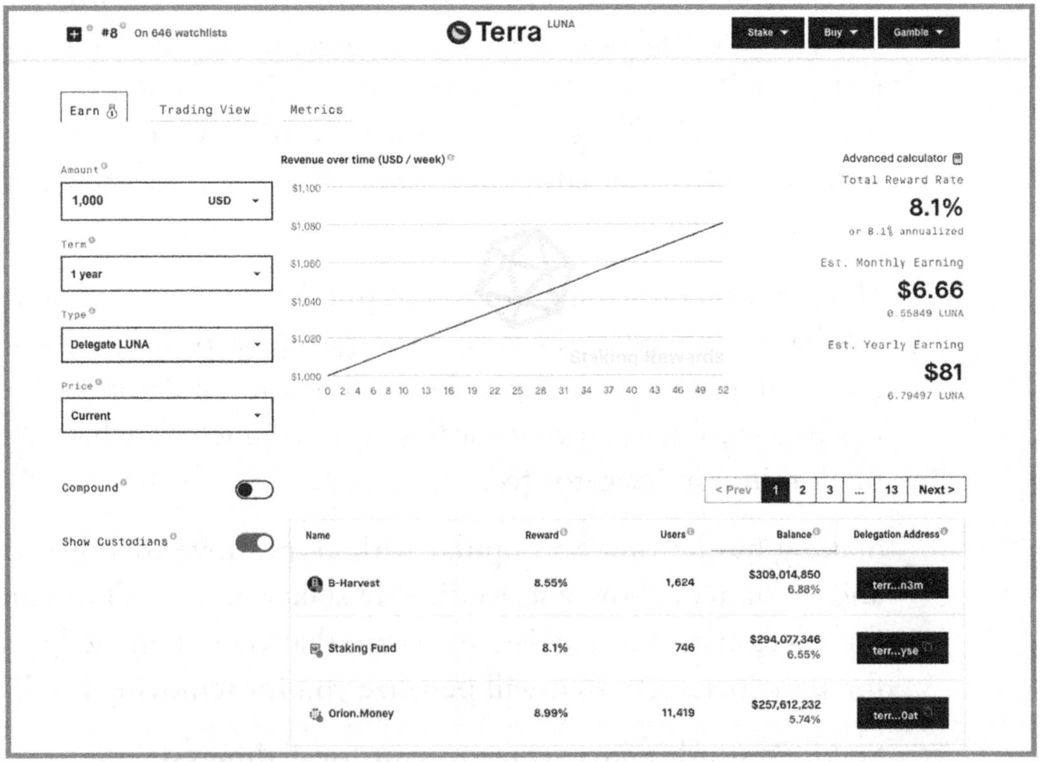

Figure 97: Terra LUNA staking rewards calculator and list of validator nodes.

Which Is Better?

New crypto investors often confuse lending and staking. There are three main differences. The first is that while staking is limited to PoS cryptocurrencies, lending is not–you can also lend out stablecoins and PoW coins. Second, PoS cryptos can yield higher rewards when staked vs. loaned. Finally, staking contributes to the security of the network, while lending does not. Investors who stake their holdings typically have more loyalty, and value their right to cast votes when it comes to network governance.

Day and swing traders who require quick access to their crypto aren't generally as interested in staking or lending. But if you plan to HODL your funds for more than a few weeks, then trying out at least one form of yield farming is a no brainer. Not only are the various options leaps and bounds ahead of legacy finance in terms of ROI, they are also straightforward to set up and fun to manage.

Part 5

Honor the Way of the Conscious Crypto Investor

Chapter 32

The Tao of Crypto

Taoism is a school of thought that originated with the Chinese mystic Lao Tzu. **The Tao** (pronounced Dao) is the way: harmony with the source of existence. Understanding the way is a lifelong process of self-discovery that involves studying the ebb and flow of Nature. Each of us is an eternal student of the Tao. When followed, it leads us to a state of humility, trust, and surrender that naturally reveals when to act and when to be passive.

What does the Tao have to do with crypto? Smart money follows the way, because going with the flow implies a minimalism that does only what is absolutely necessary. A conscious crypto investor is always thinking about how to maximize efficiency. Examples would be using trading strategies that minimize time spent staring at charts, learning how to use only the most helpful indicators, or reading only the best news sources. In the chapters that follow, we will apply the essence of the Tao to conscious crypto investing.

Jiu Jitsu: Channeling Forces

Have you ever watched a Jiu Jitsu master in a fight? Jiu Jitsu is a Japanese version of unarmed combat which allows the practitioner to respond to physical force so that it is turned against the opponent. In essence, it is the art of yielding, which in modern society seems to become more of a lost art with each passing day.

Jiu Jitsu is an example of passive action. In the cryptosphere, there is an eternal battle waged between bulls and bears that rages on 24/7, 365 days a year. The continuous activity in the market is like an aquatic ecosystem that contains every type of sea creature you can imagine. There are the little krill who like to dabble with trading small amounts of crypto after work or on the weekends. At the other extreme are high-net worth whales who wait for big moves up and down to swallow swaths of krill. Riding on the whales are barnacles: day traders who aggressively trade on very short timeframes in an attempt to scalp their peers. Institutions can either be sharks that scalp from retailers, or dolphins that absorb supply during accumulation to HODL for many years.

No matter how big of a fish you are, crypto's path of least resistance mostly involves learning when to be passive. Smart money knows how to wait on the sidelines instead of engaging in the feeding frenzy.

Lazy Lions?

If you study lions in the savanna, you will notice that they spend most of their time either sleeping or lounging in the shade. During this time they look pretty lazy, but in reality they are scoping out their next prey. They may observe a herd of zebras for days, identifying the easiest kill in order to conserve as much energy as possible. When they hunt, proper positioning and timing ensure that every movement counts.

Smart money trades the way a lion in the savanna hunts. Even when lions are ravenous, they have the discipline to wait until the most opportune time to hunt. They move with their prey, not against it. In other words, they follow the trend.

Surf coach Kale Brock once taught me that if I spent at least five minutes watching the waves from the shore before I paddled out to surf, the knowledge of how and where they were breaking would drastically increase the number of waves I caught. Crypto is no different. Instead of making impulsive trades based on the most recent activity, choose to spend weeks and months *consistently* studying charts both before and after you make your moves. Then notice the difference in how well your investments perform.

If you dedicate at least ten minutes a day to studying charts, you will learn enough in a year to dramatically improve your trades. Simply knowing how to spot a breakout vs. a breakdown is enough to give you a huge advantage over the average krill. Patient observation can lead to success in even the most volatile conditions.

Catching Falling Knives

In trading there is the expression: "Don't try to catch a falling knife." If you see the price of a coin plummet, give it some time to settle and show sideways movement before making an entry. During a downtrend, it is possible, but highly improbable to buy the bottom. If the odds are stacked against you, why take the risk?

The contrast to catching a falling knife on the way down would be "Don't try to catch a wave at its peak." A wave at its crest crashes the quickest, and with the most force. Surfers will position themselves to be lifted by a wave as it rises, to give them enough time to ride along the face of the wave. They know that if they catch the wave when it is already breaking, they are going to be sent straight down into the whitewash.

In terms of crypto, the wave is already at its peak when there is public awareness during markup. By the time dumb money realizes a frothy disaster is imminent, smart money has already taken a glossy ride along the top of the wave and off into safety. Time your entries and exits carefully to avoid catastrophe!

Discipline Is Your Greatest Ally

When it comes to investing, discipline is restraint from impulsive action, *especially* during volatile price swings. A disciplined investor is like a supple infant that can take a spill and bounce right back up again. If you yield in times of uncertainty, you will not break.

For example, let's say Tobias has seen the price of his AVAX coins double to $60 in a few short weeks, with no signs of slowing down. His FOMO impulse is to invest his entire life savings in the hope that the price will continue to moon. After all, the price of AVAX hasn't been anywhere near $30 for many months, so why would it drop back down? Sadly, Tobias is a victim of **recency bias**: making assumptions based on information on a limited timeline. *Crypto markets aren't rational,* and predicting the future based on recent events is like trying to read the mind of a chimpanzee (we are, after all, evolved primates).

If Tobias can apply discipline to calm his excited emotions and listen to the voice of reason, he will resist the allure of FOMO. Fortunately Tobias learned all about market psychology by reading this book. He chooses to take profits on his initial investment, and then sit back WOAH style and wait for a better opportunity to buy more after the price has dropped.

Discipline applies to both sizing and timing your entries. If Tobias had $6,000 to invest in bitcoin, and he knew that he would be HODLing for 10 years, how much of a difference would it make if he bought when the price of 1 BTC was $60,000 vs. $40,000? In the first case, Tobias would end up with 0.1 BTC, while in the second he would have 0.15 BTC. It may not seem like much now, but if the price of 1 BTC rises to $1 million in 5 years, that would make a $50,000 difference in Tobias' account. All from a $6,000 initial investment!

Chapter 33

Master Your Emotions

Traders like to criminalize emotion, saying that it should be eliminated altogether. But emotion is the driver sitting behind the wheel of every market. Trading without making an attempt to understand emotion or how to channel it properly is like trying to drive a car from the backseat (without self-driving).

From a broader perspective, emotions are merely energy in motion: collections of thoughts that have gathered enough momentum to grab our attention. *It is not the emotions themselves, but how you use them that will ultimately make or break you as an investor.*

A conscious crypto investor has developed a level of empathy that enables her/him to identify the dominant emotion of the collective without attaching to it. Think of an eagle that is soaring high above in the sky, passively encircling a riot of several thousand people in the streets below.

The eagle has no reason to be actively engaged in the conflict, and is simply witnessing out of sheer curiosity. It is therefore able to keep a safe distance from the madness.

On a basic level, every investor understands the inherent value of detachment. But when it comes time to put it into practice, most will fail miserably. If they didn't, market psychology would cease to exist. There would be nice smooth waves of price action all the time, with no-one jumping off a vertical cliff or clamoring up a wall.

Even traders who have been playing the game for decades make the same mistakes. Money is intimately connected to our survival instincts. When we sense that our money is at risk, it threatens our well being, and our primitive brains are wired to react with strong emotion. We are designed to have **loss aversion**: a higher likelihood of avoiding the pain of loss than taking the necessary risk to win. While this attitude may help keep you safe from a hungry tiger, it doesn't serve you as an investor. The good news is that anyone can overcome this impulse, given that they are willing to consciously rewire their brain.

Nobody gets excited by the prospect of losing money (except MrBeast). There are many financial gurus who say that your #1 mission is to not lose money, but this creates a problem. There is a ridiculously high probability that you will experience days, weeks, or months of unrealized loss, where your investments will go down. When this happens, you may feel like a failure. You may begin to doubt your decisions, thinking "I should have sold sooner," or "I wish I had never bought into this scam in the first place." And then you will start thinking about what that wise crypto sage told you, and you will feel tense and resistant. You will start to resent the game.

Remember the saint's advice? Don't play the game unless you can learn to treat gains and losses equally. No easy task, but imagine how much smoother the ride would be if you could somehow learn this mindset! Well lucky for you, losses can be minimized, and when they do occur, they can stay unrealized (you don't need to sell). The surest way to maintain your composure and to minimize loss is to be able to quickly recognize extreme greed and fear, and keep them from influencing your trading decisions.

Break Free from the Greed/Fear Cycle

Greed is the deadliest assassin lurking in the shadows of every bull market. The reason 90% of investors lose money in any market is because they buy when the price is parabolic. They buy after their best friend, his mother, and his mother's uncle's trading bot invested half their life savings. They get so wrapped up in the excitement that they begin having hallucinations of Uncle Scrooge taking his daily swim in a vault of gold coins.

The expression "pigs get fat, and hogs get slaughtered" contrasts healthy investors who take their fair share and move on to greener pastures, with hogs who will stay in a position until they become complacent. Their greed leads them meandering toward the slaughterhouse, unable to see the enormous downside that awaits them. When it all comes crashing down, they panic sell and lose everything.

Humans have the remarkable ability to let emotions beat logic into a bloody mess. Fortunately there are several external signs of extreme greed and fear that once recognized, can help us override our impulse to join the bandwagon.

Be Sharp Like an Owl

Conscious crypto investors are most at ease going against the crowd. They are always on guard against herd mentality manifesting as excessive greed and fear. Their knack for acute observation helps them to recognize emotional extremes, and quickly navigate away from them.

Here are some examples of greed:

- ✓ Multiple major media outlets and financial reports making bold price predictions. Headlines that innocently read "Is It Too Late to Buy Bitcoin?" Random celebrities are giving their seal of approval.
- ✓ A **white swan event**: an unexpected positive event or announcement that leads to parabolic rises in price. This can occur over a very short time period, and is accompanied by unusually high volume. The steeper the angle of recent price activity, the less sustainable the trend.
- ✓ Investing in crypto is suddenly cool again. Even your grandmother is advising you to buy some after she read about that Filipino on the news who bought two houses for his family from his Axie Infinity earnings.[1]

The latent greed within us is the devil lurking in the shadows. Greed will start telling you that you are a genius, and that there is no reason to sell when your investment could easily 3x from here. Even though you have witnessed others fall prey to this line of thinking, greed can create a **blind spot bias** that prevents you from seeing your own limited beliefs. The next time you become attached to your winning investment, remember these warning signs or you risk falling prey to wolves.

When greed prevails, run for the hills. Don't FOMO into an investment when public awareness is peaking. If you timed your entries properly, now is when you should be taking profits. By selling into the strength created by greed, you will be well poised to buy into the weakness of fear that follows. The more extreme the greed factor, the more you should consider exiting some or all of your position. The stronger and faster the rise, the harder the fall.

Under the right circumstances, fear can be just as self-destructive as greed. Whereas greed peaks during the euphoria of a market cycle top, fear has its place at rock bottom. It rises to power soon after decline sends the price straight down into the depths of hell. Fear is magnified when selling pressure wreaks havoc on the market for what feels like an eternity.

Retail investors who invested more than they could afford to lose are usually the first to bail out during the depressing atmosphere of a bear market. A slow, steady correction over months or years can wear on the nerves of anyone who is unprepared and impatient. Extreme fear can cause even the most seasoned investors to sell at the worst possible time.

Here are some examples of extreme fear:

✓ Sharp red declines that look like they're about to drop off the screen and into your lap. Remember that even one engulfing bearish candle after several weeks of green can be a telling sign that a reversal is rearing its ugly head. The steeper the angle and the higher the accompanying volume during a correction, the greater the fear.

- ✓ Relief rallies are brief and on lower volumes, appearing like sparse mirages in a desert of despair.
- ✓ Multiple major media outlets release an endless spew of negative press. Financial personalities and celebrities flip flop and start to trash talk what they once glorified.
- ✓ Your grandmother frantically calls you to issue a shrill warning never to invest in cryptocurrency if you hope to one day "make it in life." (cue cryptonauts retreating to their underground bunkers)

Be like a silent deer listening with its ears perked, on guard for the above warning signs. You might get away with riding bulls for awhile before jumping off into safety. But bears are a different animal, because once you see them, it is often already too late. If you are a die-hard HODLer looking at the bigger picture, then there is no cause for concern. Wolves, on the other hand, already sold into strength, and are patiently waiting to buy into weakness.

If you happen to find yourself needing to exit your positions after clearly missing the top, it is still possible to escape unscathed. After every major crypto bull run there has been a dead cat bounce. If you wait patiently for the bounce, you can sell near its peak and sit out during the ensuing bloodbath.

Never Trade When You Are Emotional

Ancient Ayurvedic wisdom says that it is better to skip a meal than to eat when upset. Can you think of a time when you were too emotional to eat? What should be a pleasant, nourishing experience can become painful when we are stressed, since the body puts our digestion on pause as we enter survival mode.

The same rules apply to trading. Strong emotions can turn even the most peaceful monk into a jittery clown. You are no exception! For this reason, you should ask yourself the following questions before making any trade:

1. How am I feeling right now? Am I balanced mentally and emotionally, or am I overstimulated and exhausted after a long day? Am I simply hoping for a quick dopamine boost? Are there any traces of FOMO, or am I more aligned with WOAH?
2. What is the current sentiment of this coin? Is greed or fear predominant?
3. Is this a logical trade, one that is based on a clear analysis that minimizes my risk and maximizes my reward?

Remember being told to think before you act? Here it's not just your reputation on the playground at stake–it's your hard earned money. Without clarity of mind, you can make the wrong decisions no matter which direction the price is moving.

Control Emotion with Sacred Pauses

A **sacred pause** is a moment of pure awareness. It isn't about doing, it is about being. A sacred pause is whatever you do to remember your eagle perspective. It can happen during a series of deep breaths, a short hike in nature, or while watching the sunset. In the same way that you'll never regret brushing your teeth, sacred pauses can only enhance your mental state. Taking sacred pauses throughout the day–even if only for a couple of minutes at a time–can gradually free you from extreme emotions. *One who has mastered emotions cannot be enslaved by uncontrollable market conditions.*

Meditation is the ultimate sacred pause. As a conscious crypto investor, meditation is a practice that will dramatically increase your discernment. It will also help you to develop your intuition, so that you can make important decisions with split-second precision. Meditation is the gift that will reward you exponentially as you progress from one day to the next. You can meditate anywhere, at any time of the day, for any duration, and you will *always* benefit.

It is normal to feel resistance toward the practice of controlling your mind through meditation. Your mind is already running in a hundred directions at once, why would it want to be leashed? But take a moment to think about meditation in terms of money. In a volatile market where one wrong decision can cost you a fortune, wouldn't you want to have as much of an edge as possible? Don't you want to do whatever it takes to increase the chances of making the right decisions? I know I do, which is why my meditation practice is paramount.

The best time to meditate is right when you wake up in the morning. Instead of grabbing your phone to glance at your crypto portfolio, see if you can sit upright and close your eyes for ten minutes, focusing on breathing as deeply as possible. Inhale and feel the stomach expand, and then exhale as it slowly draws back in again. Treat the thoughts like a passing dream, entering the river of existence and floating away as quickly as they came.

Challenge yourself by practicing for 21 days straight. If you skip or forget one day, start again. Consistency creates momentum. Your overall awareness will begin to expand. You will have more discipline, more calm and clarity, and more vital energy available throughout the day. You will feel less of a victim to the crazy volatility of the cryptosphere. Most importantly, you will begin to act from a place of inner knowing. As you expand, your thoughts, words, and actions will shift to align with your abundant nature.

Meditation isn't New Age hocus pocus. It is the ancient science of mind that has been passed down by great sages for millenia. Investing is a head game, and if you aren't constantly sharpening your mind, then you will soon find yourself among a pack of hungry wolves.

Chapter 34

Be Like Low-Lying Grass

Have you ever witnessed a cataclysmic event such as a tornado or an earthquake? They have the power to cause massive destruction in a matter of minutes. If you pay close attention, you will notice that the largest, most rigid objects–those that appear strong on the surface–are the quickest to be annihilated. Towering skyscrapers crumble to the earth, giant trees are uprooted, and semi-trucks are tossed around like toys.

But then if you look at the ground during a catastrophe, you'll notice that the grass is unphased, and the rivers never stop flowing. No matter how many times we trample the grass underfoot, it still somehow springs back up again. And no matter how big the boulders are that stubbornly sit in the river, it always manages to find a way around them.

Humility should be the very foundation of investing. Humility is the source of great strength during difficult market swings. If you can restrain yourself from thinking you are the master of the universe when your investments are up 5x, then you will be more likely to take profits at the right time. Similarly, if you can keep your sanity and HODL when **paper hands** are panic selling their coins during a crash, you will survive.

Humility also makes it easier to have an "easy come, easy go" mindset. Losses won't phase you as much, and gains won't inflate you either. Humility makes you patient, resilient, and receptive. These are all qualities that keep you on the leading edge, supporting an inner vibration that will naturally pull you toward the most lucrative investments.

Have you ever watched bamboo swaying in strong winds? The bamboo seems fragile on the surface–it is very tall and lanky, and inside it is hollow. But its roots run deep, and it knows how to flex under pressure. The strength of bamboo lies in its ability to yield. Investors who are able to yield are considered to have **diamond hands**: mental strength to not sell their coins during the winds of volatility.

Channel Smart Money

Smart money is always devising new ways to stretch outside familiar comfort zones. Constant self-analysis keeps conscious crypto investors on guard against narrow minded thinking. The expression "know thyself" is a mantra that they adhere to in order to prevent **confirmation bias**: a form of self-deception where one is drawn to information that proves existing thoughts, while conveniently ignoring contradictions.

As an example, you may notice that you tend to favor either bullish or bearish crypto influencers depending on how much money you have invested. When you are heavy in cash, it is easy to side with bears, who may be predicting another 50% dip. But if you find yourself feeling overextended during a stab wound correction, you are much more likely to favor bulls who dismiss it as a minor hiccup that will soon resolve.

An honest self-awareness is the cure for confirmation bias. Always be willing to entertain all possible scenarios, and resist the tendency to imprison yourself inside echo chambers where others mimic your beliefs. Be careful not to immediately dismiss those with conflicting views, and have the flexibility to quickly change your mind as soon as you realize you are wrong.

The more receptive we become to the views of others–regardless of their experience or credentials–the more tools we will have to make an expansive analysis that is free from the clutches of confirmation bias. We should always try to remain innocent of **anchoring**: jumping to conclusions based on the first piece of information we discover.

The famous investor J.P. Morgan is known to have said, "Millionaires don't use astrology, billionaires do."[1] This comes as a shock to many traditional investors, who know him as a banker and financier. But according to Astrologist Evangeline Adams in her 1926 autobiography, The Bowl of Heaven, Mr. Morgan had an avid interest in astrology and its effects on the stock market, politics, and business. In his later years, he was a regular subscriber to her reports, and even invited her to join him and his party on his yacht in the Far East to "spend several months on a scientific investigation of the occult."

It should come as no surprise that J.P. Morgan had a personal astrologer. In fact, he had an entire team of advisors helping to inform his investment decisions. If you had as much capital at stake as he did, wouldn't you too?

The trick is to always act as if there are billions of dollars on the line with every choice you make. This mindset is what drives smart money to constantly seek novel and unconventional approaches to investing. When the pressure is high, you naturally become more thorough and intentional in your decision making process.

One example of an alternative approach to investing would be using the "Moon Phases" indicator in TradingView to track changes in asset prices over the seasons. Another would be studying weekend trends, or investigating price patterns during certain times of the month or year. There are also many different ways to apply Fibonacci indicators to determine levels of support and resistance. Be like a detective that is searching for patterns of continuity in a seemingly chaotic market. Oftentimes it is the most far fetched ideas, such as the Elliott Wave theory and Wyckoff method, that eventually receive widespread adoption (hint hint).

Keep the Numbers to Yourself

If you clearly understand the essential principles laid out in this book, you will be on the leading edge of the cryptosphere. Your timing will be impeccable, and your gains will be exponential. In fact, you will be so spot on with your investments that friends and family will think you are on a lucky spree. When your "luck" lasts years or even decades, there is a strong likelihood that they will either do one of two things: 1. curse you for your prosperity behind your back, or 2. praise you for making sound investments, and seek your counsel.

You quickly learn who your true friends are when you become successful and prosperous. Money is an amplifier that can bring out both the best and worst in people. Others are going to judge you no matter how much money you have, and you shouldn't stop trusting those who have shown themselves to be trustworthy. Still, it is always necessary to exercise caution.

You can freely share investment tips and strategies with those who express interest. By doing so, you are not only spreading an abundance mindset, but also refining your investing outlook. You can even share which coins you hold in your portfolio and their percentage weight. But you should never share the actual number of coins or dollar equivalents.

It is also unwise to share public addresses where you conduct many transactions. Instead, consider creating alternate addresses that are strictly for receiving crypto. Even sharing how much money you made last year with close friends and family can be risky. Gossip is human nature, and conscious crypto investors don't want to be entangled in any unnecessary use of energy. While this may sound self-explanatory, you'd be surprised just how much your ego will tempt you to shout your successes from the rooftops. Check out your instagram or facebook feed for a daily dose of evidence (#DeleteInstandFacebook).

Exist Outside Conditions

This is really quite simple...that is, in the absence of that gigantic ego that clings to us like a rabid monkey. When the price is going up, it has nothing to do with you. If you begin to feel proud, smart, or ecstatic, observe and adjust. Maintain your frame like a warrior on the battlefield. If the warrior becomes too proud through identification with small victories, it is a weakness that can easily blind him/her to other dangers lurking in the shadows.

One of the biggest mistakes investors make is to become complacent, carelessly letting their greed run wild. Always check yourself before you wreck yourself. *If you don't take profits off the table, the market will do it for you.*

When the price is going down, it also has nothing to do with you. If you begin to feel worthless, stupid, or depressed, observe and adjust. Getting down on ourselves because the market is down is like crying whenever it rains. It's going to happen, and whether you laugh or cry, days will pass by. We should strive to feel good all the time, regardless of whether the market is pumping or dumping. Longevity in the cryptosphere comes from developing the wisdom to keep clear from either over-inflating or deflating ourselves under any circumstances.

Chapter 35

Chill Out and Go to Recess

Remember being a kid in school, waiting desperately for 2nd period to end so you could catch some fresh air and toss a ball around? It felt like it was barely two minutes, but it gave your mind a rest from the onslaught of useless information. Schools offer breaks at key intervals throughout the day because without recess, even the most chill students would be flipping tables.

Every day in the cryptosphere there is some new sensational headline, or some coin that just pumped 40%. When you give crypto too much of your focus, it becomes easy to get wrapped up in all the hype during the euphoric highs of a bull run. It's also just as easy to fall into a black hole of depression when the PTB decide to crucify crypto for the 12th week straight.

Our minds have limits that we need to recognize and respect in order to maintain health, happiness, and mental well being. To save you from being sleep deprived and riddled with anxiety, stick to your crypto routine. Practice limiting yourself, so that you can enjoy recess as much as possible.

Know When to Detach

Can you guess the worst times to be thinking about crypto? First thing in the morning, and right before bed. These are regenerative hours, when you should be disengaged from material concerns. For the sake of a peaceful morning, and a good night's sleep, during these times you should resist the temptation to look at anything crypto. We create enough tension in our minds thinking about money throughout the rest of the day.

Your mind is incredibly powerful. When you first open your eyes in the morning, where you put your focus is setting the tone for the day. If you see your portfolio is down 30% since last night and then get sucked into a FUD storm, are you increasing or decreasing the chances that you'll have a good day? Even if you see your portfolio is up, do you *really* need to know about it right away? Follow the **golden rule of investing**: your well being is your top priority. Invest in yourself first.

For some of us, having the discipline to keep the phone more than an arm's length away sounds impossible. Sure, it seems harmless to take a peek at your phone for half a minute a few times during regenerative hours. Maybe the first five times it is harmless. Then when it becomes a habit without you realizing it, and your screen time is up 20% from last week, it's already too late!

Keep the obsession in check by setting a strict schedule. On the iphone, for instance, you can set a daily bedtime reminder a couple of hours before you actually go to sleep. Then when the time comes, you can put your phone in airplane mode with the intention of turning it back on *after* you've taken at least some time to collect your thoughts, envision a joyfully productive day, and groom yourself.

Skip the Weekends

In a market that never sleeps, the weekends aren't spared from volatility. Manic buys and panic sells aren't uncommon on Saturdays and Sundays, especially after a quick run up or down. There are three main reasons:

1. Professional crypto traders who previously spent decades trading stocks, bonds, and commodities aren't available on the weekends. They are used to closing out positions on Friday, and enjoying a mental reset over the weekend. This is important to understand, because this caliber of trader helps to enhance the stability of the market during the week.

2. Institutions are also sidelined on the weekend, since banks are closed. Investment firms offer large amounts of liquidity to the market during the week, but on the weekends they aren't able to move fiat between exchanges to buffer the order books. Less liquidity leads to greater spreads, and a higher likelihood for volatility.

3. Less experienced retail krill, most of whom work full time jobs during the week, are typically more active on the weekends. To make matters worse, people tend to consume more alcohol on the weekends, which we all know lowers inhibition. Herd mentality reigns.

These factors add an element of complexity to the equation that you won't have to deal with in traditional markets. At this stage in crypto adoption, we already have to put up with enough volatility. There is no need to add any more stress to your life by trading on the weekends when you should be spending quality time with your family.

If you've zoomed out and taken a long term approach, you won't be phased by weekend swings. More importantly, your entries and exits won't occur during all the commotion. Dramatic sell-offs can quickly rebound by the time the banks open on Monday morning. Keep this in mind when the impulse to jump ship arises!

Chapter 36

The Conscious Crypto Investor

Forget every trading decision you've ever made. Forget how much money you made or lost. None of it matters anymore. The expression, "don't cheat on your future with your past" is sage counsel for the conscious crypto investor.

There are hundreds of thousands of investors who lost money in the last bear market. I know one in particular that lost a fortune in 2017, and now he can't seem to stay in trades for more than a day or two. When I mention a new DeFi project that sounds promising, he defaults to cynicism, saying it will likely tank like all the other 2017 ICOs. He's completely paranoid.

We've all made decisions in life that have caused mental and emotional wounds. If we maintain a beginner's mind at all times, every experience becomes a valuable life lesson. If you can anticipate potential losses without feeling anxiety, then when it happens, it won't destroy your peace of mind.

Your mistakes will teach you more than any book or course ever will. When you make a mistake, have the conviction that it will save you piles of grief in the decades to come. Go easy on yourself, and remember that both victories and defeats are stepping stones toward financial freedom.

I remember the first time my son asked me for money, and I gave him a handful of coins. He was four years old, and he was thrilled to hear the clinking sound the coins made in his pocket. Two minutes later, he groaned as he shoved his fist into his pocket and pulled out the change, handing it back to me.

"What's wrong?" I asked him.

"Daddy it's *heavy*! Will you please take it, I don't want it anymore!"

I laughed, and he ran off to play with his toys.

When all is said and done, money is simply a language to express value–how much you have doesn't determine *your* value as a human being.

For Generations to Come

As you invest in crypto, you should always be aware that your decisions will impact your family for several generations. Crypto accounts will become the trust funds of the future. As legendary investor Michael Saylor is fond of saying, investing in bitcoin today is like buying a block in Manhattan over a century ago. Since crypto will continue to appreciate exponentially, you want to be able to pass it on to support your family for as long as possible.

Keeping this in mind, prioritize carefully training your benefactors about cryptocurrency. In a matter of 30-60 minutes, you can show them how to locate and decipher your seed phrase(s), access your account passwords, and more. Ideally you teach them everything you learn in this book, since these are invaluable financial skills that every human being will one day need to master. Writing checks and visiting bank tellers are nearing their inevitable demise. The oldest, biggest banks will one day become museums visited by children as they learn about the evolution of cryptocurrency.

Traits of a Conscious Crypto Investor

Conscious crypto investors who succeed in the long run don't have to be wizards at TA, or stone cold traders without a trace of emotion. They are ordinary people who are calm and collected, patiently watching the markets with a sense of detachment. They lean away from extremism, and choose to be lighthearted in the face of adversity.

A conscious crypto investor knows how to strike a balance between investing in others and dedicating time, money, and energy for him or herself. He understands that since the crypto market is completely outside his control, it is unwise to let it be the focus of all his resources and attention. He strives to maintain clarity about exactly how much time, money, and energy he is willing to put into the cryptosphere.

Like water, a conscious crypto investor is always seeking to flow with effortless ease. She intends to be increasingly mindful of her thoughts and actions, to simplify the process of investing so that it can bring ultimate joy and fulfillment. She does this gradually, through continuous study and a commitment to a sound investment strategy.

A conscious crypto investor casts his vote for projects with genuine utility that are enhancing lives. He chooses to focus only on groundbreaking projects that are both meaningful and necessary. He takes the time to gather information about their underlying technologies, study their ecosystems, track their charts, and more.

A conscious crypto investor is patient and persevering under all circumstances. She possesses an open mind, and is willing to take the necessary risks to break free from the legacy financial system. She revels in the opportunity to reclaim her sovereignty by personally becoming her own bank.

Thank you for choosing to join the ranks of millions of conscious crypto investors who are actively securing their financial future. Whether you are new to investing or are a seasoned veteran, it is a great joy and privilege to be your comrade in the cryptosphere. May you use your knowledge and wisdom to empower others to take your lead, for the benefit of many generations to come.

> "Down goes the tough and big. Up comes the tender sprig."
>
> – Lao Tzu

Want leading edge information on supercar cryptos?

With crypto annual adoption nearly double what internet adoption was in the 90's, the cryptosphere moves *fast*. Imagine the difference having access to the most relevant and essential information can make on your portfolio. How much time and energy you could save by receiving daily insights from a community of insiders who openly share a wealth of information, including:

- In-depth TA for all of the coins in our conscious crypto portfolio. Learn how to buy and sell at the most advantageous times.
- Top picks across every major crypto category. Know about lower cap gems *before* everyone else.
- Exclusive video courses that help to walk you through the basics of getting started, and expand on the most cutting edge concepts in this book.
- Weekly newsletters that include altcoin reviews and important news that will help you to make educated decisions you won't regret later.

If you are interested in joining a thriving community of conscious crypto investors, and want to learn about the hottest coins that can lead to explosive returns, join the conscious crypto collective at defimagick.com

Appendix A: References

Chapter 2

1. https://www.treasury.gov/resource-center/faqs/currency/pages/legal-tender.aspx
2. https://www.somagnews.com/9-trillion-story-22-of-us-dollars-printed-in-2020/
3. https://cointelegraph.com/news/us-printed-more-money-in-one-month-than-in-two-centuries

Chapter 3

1. https://www.federalreserve.gov/monetarypolicy/reservereq.htm
2. https://en.wikipedia.org/wiki/List_of_bank_runs
3. https://www.visualcapitalist.com/purchasing-power-of-the-u-s-dollar-over-time/

Chapter 4

1. https://www.internetworldstats.com/emarketing.htm
2. https://cryptopotato.com/growing-faster-than-the-internet-bitcoin-projected-to-have-1-billion-users-by-2025-analyst-says/
3. https://cointelegraph.com/news/crypto-population-doubled-to-over-200m-users-since-january-report-says
4. https://news.bitcoin.com/bitcoin-history-part-14-the-1000-btc-poker-game/

Chapter 5

1. https://www.dallasnews.com/opinion/commentary/2021/08/15/50-years-ago-the-us-closed-its-gold-window-triggering-the-failure-of-our-monetary-system/
2. https://blog.chainalysis.com/reports/2021-global-crypto-adoption-index
3. https://www2.deloitte.com/content/dam/insights/articles/US144337_Blockchain-survey/DI_Blockchain-survey.pdf
4. https://www.statista.com/statistics/268225/countries-with-the-highest-inflation-rate/
5. https://bitcoinminingcouncil.com/wp-content/uploads/2021/07/2021.07.01-Mining-Council-Press-Release-Q2.pdf
6. https://www.coindesk.com/policy/2021/10/01/el-salvador-mines-first-bitcoin-with-volcanic-energy/

Chapter 6

1. https://globalfindex.worldbank.org/sites/globalfindex/files/chapters/2017%20Findex%20full%20report_chapter2.pdf
2. https://www.worldbank.org/en/news/press-release/2021/05/12/defying-predictions-remittance-flows-remain-strong-during-covid-19-crisis
3. https://www.yahoo.com/now/countries-using-cryptocurrency-most-210011742.html

Chapter 7

1. https://www.wsj.com/articles/verizon-to-buy-aol-for-4-4-billion-1431428458

Chapter 13

1. https://www.statista.com/topics/1551/online-gaming/
2. https://decrypt.co/77949/ethereum-game-axie-infinity-daily-users-10x-1-million
3. https://www.forbes.com/sites/tjmccue/2020/02/05/looking-deep-into-the-state-of-online-video-for-2020/
4. https://www.sandvine.com/blog/youtube-rules-mobile-2019-mobile-internet-phenomena-report-released
5. https://cointelegraph.com/news/the-irs-offers-a-625-000-bounty-to-anyone-who-can-break-monero-and-lightning
6. https://research.thetie.io/musk-and-saylor-tweets/

Chapter 30

1. https://www.latimes.com/california/story/2021-07-26/judge-rules-against-fbi-beverly-hills-raid

Chapter 33

1. https://cointelegraph.com/news/axie-infinity-player-buys-two-houses-in-the-philippines-from-in-game-profits

Chapter 34

1. https://www.forbes.com/sites/kenrapoza/2012/02/20/can-planets-affect-your-portfolio/?sh=de8b67615e16

Appendix B: Recommended Resources

Cryptocurrency key metrics: coingecko.com, coinmarketcap.com, messari.io

Cryptocurrency news: cointelegraph.com, coindesk.com, decrypt.com, cryptopotato.com, cryptopanic.com

Blockchain explorers: blockchain.com, etherscan.io, bscscan.com

Best DEXes: thorswap.finance, uniswap.org, sushi.com, pancakeswap.finance, injective.exchange

Best CEXes: ftx.com, kucoin.com, kraken.com, binance.com

Smart trading platform: 3commas.io

Ethereum gas tracker: etherscan.io/gastracker

Exchange fee comparisons: cryptofeesaver.com/exchanges/fees

Best software wallets: Trust wallet, Exodus wallet, Coinomi wallet

Best browser extension wallets: MetaMask, Xdefi, Terra Station (for LUNA)

Best hardware wallets: ledger.com, trezor.io

Best VPNs: expressvpn.com, nordvpn.com, protonvpn.com

Chart analysis: tradingview.com

On-chain metrics: whalemap.io, glassnode.com, cryptoquant.com

Sentiment analysis: alternative.me/crypto/fear-and-greed-index/, augmento.ai, lunarcrush.com

Alt-season index: blockchaincenter.net/altcoin-season-index/

YouTube crypto influencers: Coin Bureau, Ellio Trades Crypto, Altcoin Daily, Into the Cryptoverse, Coinskid

Crypto subreddit: reddit.com/r/CryptoCurrency/

Standalone password managers: lastpass.com, 1password.com

Authy 2FA guide: authy.com/guides/

DeFi lending rates: defirate.com/lend/

Decentralized lending platforms: anchorprotocol.com, aave.com, compound.finance

Centralized lending platforms: nexo.io, celsius.network, blockfi.com

DeFI TVL: defillama.com

Staking rewards: stakingrewards.com

ETH Flippening watch: buybitcoinworldwide.com/flippening/

Bitcoin halving countdown: buybitcoinworldwide.com/bitcoin-clock/

Cryptocurrency project calendar: coinmarketcal.com/en/

Find your nearest bitcoin ATM (BTM): CoinATMRadar app

Portfolio tracker apps: coinstats, cointracker, coinmarketcap

Crypto performance comparisons: cryptobubbles.net

Bitcoin obituaries: 99bitcoins.com/bitcoin-obituaries/

P2P bitcoin transactions: paxful.com, localbitcoins.com

Convert satoshis to bitcoin: finder.com/satoshi-to-bitcoin-conversion-calculator

Store twitter threads: readwise.io

NFT marketplaces: rarible.com, opensea.io, mintable.app

Bitcoin mining energy statistics: bitcoinminingcouncil.com

Glossary

Accumulation: The first stage of a market cycle, which occurs after decline. A period of consolidation where smart money buys in increments.

Active wallets: The number of wallets with a crypto balance. An on-chain analysis measure of network growth.

All time high (ATH): The highest price a cryptocurrency has traded at throughout its market history.

Alt-season: When 75% of the top fifty altcoins performed better than bitcoin over the past ninety days.

Altcoin (alt): Any alternative cryptocurrency to bitcoin.

Altcoin Index: A popular website that tracks the top fifty altcoins by price performance relative to bitcoin. It is an excellent place to search for strong investment candidates.

Anchoring: A specific bias that involves jumping to conclusions based on the first piece of information presented.

Aping: Impulsively buying large amounts of cryptocurrency without paying attention to market conditions or the voice of reason. Usually a symptom of intense FOMO...or the aftermath of a strange craving for bananas.

Ask price: The price on the books at which sellers are willing to sell their coins.

Automated market maker (AMM): DeFi protocols that use mathematical formulas to set token prices. Smart contracts are used in place of buy and sell orders to link buyers and sellers. AMMs are used on decentralized exchanges.

Bad news bears: People who spin crypto conspiracy theories that make even the most die-hard enthusiasts want to run for cover.

Bank reserve: A specific amount of capital that you choose to keep in the bank under all circumstances. By keeping a bank reserve, you won't stress as much if and when your portfolio takes a hit. You will also have a cushion of funds to cover expenses if there is a prolonged bear market.

Bear market (bear run): An extended downtrend that can last many months or years. Bears swipe their claws down. The entire cryptosphere experiences a slow, painful bleed out during the cold winter of a bear market.

Bearish divergence: When the price is moving higher while other technical indicators are moving lower. An example would be the price increasing over time while the volume is steadily decreasing.

Bearish reversal: A trend reversal that happens after a breakdown below an uptrend line. See also bullish reversal.

BEP-2, BEP-20: Tokens that follow protocols on the Binance smart chain.

Bid price: The price on the books at which different buyers are willing to purchase coins. Highlighted in green.

Big three pillars of analysis: When taken together, fundamental analysis, sentiment analysis, and technical analysis will give an expansive view of a project that helps conscious crypto investors to make the most informed decisions possible.

Bitcoin: The original cryptocurrency developed in 2008 by Satoshi Nakamoto. Bitcoin is a truly peer-to-peer, decentralized digital payment network based on cryptographic proof. Bitcoin is capitalized to refer to the network, but lowercase as a currency.

Bitcoin Bull & Bear Index: A sentiment analysis tool that uses AI to track social media bitcoin conversations on twitter, reddit, and bitcointalk. A ranking of 1 is extreme bullish, while 0 is extreme bearish.

Bitcoin dominance: A percentage that shows the relative market cap of bitcoin compared to the rest of the cryptosphere.

Black swan event: A negative event or announcement that catches investors off guard, increasing sell pressure and leading to a systematic correction. During these times the price will unexpectedly drop 50% or more. The initial response to the spread of COVID-19 created a black swan event. See also white swan event.

Blind spot bias: An inability to see one's own biases despite recognizing those of others.

Blockchain: A chain of blocks containing data that is hashed to prevent them from being altered.

Blockchain explorers: Search engines for data stored on the blockchain, which are often used to track the status of transactions.

Blow off the top: A sharp stab-wound correction that follows a parabolic rise in the price of a cryptocurrency, and signals the beginning of a bear market.

Blue-chip cryptos: Top 20 cryptocurrencies, which are generally considered to be more fundamentally sound. They have often been around longer, and are reputable projects with a greater use case.

Bouncing ball effect: Used to describe price action during a strong downtrend. Each time the price drops to a lower low, some of its force dissipates as it hits the bottom, so that it bounces back up with less momentum. This creates a series of lower highs and lower lows.

Brave browser: A standard free web browser that eliminates ads and ad trackers, creating three times faster load times. Brave won't sell your data to third parties, and their advertising model is opt-in and privacy centric. Users can earn Basic Attention Tokens (BAT) for viewing advertisements.

Breakdown: Price movements down and away from a key level of support.

Breakout: Price movements up and away from a key level of resistance.

BTM: Acronym for Bitcoin ATM, where you can use cash to buy bitcoin and send it to a wallet using QR codes. BTMs also allow you to purchase several other cryptocurrencies, and to sell them for cash. Fees and requirements differ by country.

Buffer zone: A 1-5% cushion for the price to move either above or below a trendline without invalidating it. Buffer zones provide some mental flexibility in the event that a departure from a trendline is only temporary.

Bull market (bull run): An extended uptrend that can last many months or years. Bulls buck their horns up.

Bull market support band: A helpful TradingView indicator that includes both the 21 week EMA and 20 week SMA. Prices above the band support a bull market, while prices below signal a bear market.

Bull trap: A temporary relief rally where bulls buy the dip thinking the price will recover relatively quickly, only to be overcome by bears driving the price further down.

Bullish divergence: When the price is moving lower while a technical indicator is moving higher. An example would be the price decreasing while the momentum begins to increase.

Bullish reversal: A trend reversal that happens after a breakout above a downtrend line. See also bearish reversal.

Buyback-and-burn: When a crypto project buys back a portion of the circulating supply and removes it from circulation. This is an irreversible transaction that can be viewed on a blockchain explorer, and increases the price of existing coins by reducing the total supply. See also token burn.

Buying into weakness: Buying part or all of a position as the price of an asset is decreasing.

Candlesticks (candles): Originally created by Japanese farmers to track the price of rice, candlesticks are visual displays of opening and closing prices, including highs and lows for the specified time period. The wick/shadow on the top and bottom of the candle reveal the highest and lowest prices for the time period.

Capitulation: An event that occurs when the last of the weak hands finally give in to the pressure and sell during a systemic correction, and the price bottoms out. It can usually be identified as a long lower wick in a bear market. See also weak hands.

CBDC: An acronym for **C**entral **B**ank **D**igital **C**urrency. CBDCs are digital versions of fiat currencies that will allow for unprecedented levels of government surveillance and control. Not to be mistaken with the trustless, permissionless, immutable, borderless features that are hallmarks of all truly decentralized cryptocurrencies.

Center point: The place of grounded awareness that helps us to make smart investment decisions.

Centralized exchange (CEX): A crypto trading platform that behaves similar to a traditional brokerage, since an intermediary is used to help you trade. You pay fees that enable the CEX to cover its operating expenses, to provide liquidity for trading, and to store your crypto.

Coins: Cryptocurrencies for standalone blockchain networks such as Ethereum, Cardano, and Solana.

Cold wallet: See hardware wallet.

Collateralized Debt Position (CDP): Locking up collateral in a lending protocol's smart contracts allows users to withdraw a percentage in the form of a stablecoin loan. The amount of collateral locked in a CDP needs to exceed the value of the loan by a percentage set by the lender.

Confirmation bias: A form of self-deception where one is drawn to information that proves existing thoughts, while conveniently ignoring contradictions.

Confluence: A comprehensive trading strategy that uses multiple indicators and signals to come to conclusions. For example, high volume accompanying a steady rise in price would create confluence for continuation of an existing uptrend.

Conscious crypto investors: Brave early adopters who aren't afraid to expand their horizons to both discover and invest in groundbreaking cryptocurrencies. Someone who is not only committed to steady financial growth to support one's own family, but also approaches investing as a spiritual endeavor that will ripple out to the larger global community.

Consolidation: See sideways trend.

Correction: When the price of a cryptocurrency drops 10% or more. Corrections can last anywhere from a few hours to days, weeks, months, or years. Small corrections are like scratches, medium corrections are like cuts, and large corrections are like stab wounds for a cryptocurrency.

Crypto Fear & Greed Index: A sentiment analysis tool for bitcoin that combines volatility, momentum, social media interactions, bitcoin dominance, and google search trends to give a numerical ranking on a scale of 0-100 between extreme fear and extreme greed.

Crypto supercars: Cryptocurrencies that outperform their rivals, yielding the highest annual gains.

Crypto Twitter (CT): The same twitter, just more fancy lingo and a borderline psychotic passive aggressive vibe that confuses/alienates nocoiners/normies/newbs. Honestly though, it can be a good resource...sometimes.

Cryptocurrency: A unique form of digital currency that is secured by cryptographic code. Cryptocurrency requires the internet for its storage, security, and transmission. Bitcoin was the first cryptocurrency.

Cryptocurrency Coins: Units of currency on standalone blockchain networks such as Ethereum, Solana, and Luna.

Cryptocurrency Tokens: A specific type of cryptocurrency coin. Tokens are units of currency that are used on existing blockchain networks. Aave, for instance, is an ERC-20 token, since it follows specific protocols on Ethereum's network.

Cryptonauts: Crypto users who spend ungodly amounts of time learning about crypto by reading this book (and recommending it to their friends).

Cryptosphere: A term that refers to the entire cryptocurrency space, which has the qualities of an interconnected sphere.

Custodial wallet: A wallet/keychain that can be accessed by an outside authority such as an exchange. See also non-custodial wallet, keychain, wallet.

Dead cat bounce: A bull trap that happens after a blow off the top, where the price spikes back up temporarily after a sharp decline. This gruesome analogy is given because it seems to be a knee-jerk response for so many bulls that never fails to play out, even after the most parabolic rises.

Death cross: When a short term moving average crosses a longer moving average on the way down. An example would be the 50 day EMA crossing the 200 day EMA.

Decentralized Application (Dapp): A combination of multiple smart contracts, which can be imagined as an interconnected web of money legos on the blockchain. Dapps are open-source, borderless, and permissionless. There are Dapps for trading, lending, borrowing, and gambling.

Decentralized exchange (DEX): Dapps that use smart contracts to create Automated Market Makers, where the users themselves can pool their coins to earn a percentage of transaction fees as liquidity providers. DEXes are non-custodial, since they operate in a trustless environment.

Decentralized Finance (DeFi): A decentralized network of financial applications that run on the blockchain. Defi creates the digital infrastructure that allows for the seamless storage and transfer of assets using Dapps.

Decipher code: An alphanumeric series that will help you to remember how you rearranged the order of your seed phrase. A manual encryption device that is meant to be used as an added layer of security.

Decline: The fourth and final stage of a market cycle, when bears regain control as the price plummets below previous support. A clear downtrend is formed, with lower highs followed by lower lows, and moving averages that slope down.

Decoupling: In cryptocurrency, the idea that the price of an altcoin(s) will no longer depend on the price movements of bitcoin. Ex: "ATOM will decouple from bitcoin."

Delegator: A crypto investor who agrees to stake a portion of his/her crypto holdings to a validator node in return for a percentage of the fees earned.

Diamond hands: A reference to investors who have the mental tenacity to HODL their investments during systematic corrections.

Distribution: Stage three of a market cycle, which begins as bulls and bears vie for control. Prices may continue to climb higher despite decreasing volume and price momentum, which can trick many unsuspecting investors into complacency. Smart money is slowly but surely selling their bags during distribution!

Doji: Japanese for mistake. Dojis form when the opening and closing prices of an asset are nearly identical, creating a small body that often has a long upper and lower wick. Dojis resemble crosses.

Dollar cost averaging (DCA): The practice of gradually investing in one or more assets over weeks, months, or years. See also scaling out.

Doubling down: Buying twice as much crypto with each successive entry as the price dips.

Downtrend: A price trend characterized by both lower highs and lower lows as the price steadily decreases.

Dumb Money: An impulsive attitude that can be seen in those who are often scattered in their thinking and extreme in their emotions. A way of being that leads one to act before thinking, making uninformed investment decisions that maximize risk and the potential for loss. It is common among gamblers and addictive personality types, though we all have the potential to be dumb money.

DYOR: An acronym for **D**o **Y**our **O**wn **R**esearch. The crypto version of due diligence. Conducting an in-depth investigation into a project before making decisions, to ensure that it is a sound investment.

Engulfing Bearish Candle: A red candle that is large enough for the preceding green candle to fit inside.

Engulfing Bullish Candle: A green candle that is large enough for the preceding red candle to fit inside.

Entry: Opening a position by purchasing a cryptocurrency.

ERC-20: Tokens that follow protocols on Ethereum's network.

Evening Star: A specific type of doji that appears after a considerable uptick in price. See also doji.

Exchange Traded Fund (ETF): A group of securities, such as bonds, stocks, or commodities, that trade on a public exchange. Bitcoin ETFs will allow traditional investors to access cryptocurrency through legacy financial brokerages.

Exit: Closing a position by selling a cryptocurrency.

Exit scam: When a project founder(s) disappear with money raised during an ICO or NFT sale. See also Initial Coin Offering.

Exponential moving average (EMA): A special type of moving average that gives more weight to recent price data, which means it won't have as much of a lag as a simple moving average (SMA). See also moving average.

Fair launch: A cryptocurrency with coins that are mined gradually over time using Proof of Work. None are allocated to investors in advance.

Fakeout: When the price temporarily breaks above or below a trendline, then suddenly pulls back to resume its normal course.

Fear of Missing Out (FOMO): When retail investors feel the strong impulse to buy a cryptocurrency that is pumping out of fear that they may never see the opportunity again. A form of herd mentality, or "jumping on the bandwagon."

Federal Reserve (Fed): An independent central bank that can make decisions without approval from the federal government. Since its creation in 1913, inflation has increased over 2,500%.

Flash Sale: An opportunistic view of a sharp, fast correction. Savvy investors who have been waiting on the sidelines, holding a reserve fund of dry powder, will buy coins at discount prices during flash sales. Also known as a fire sale.

Fractional reserve banking: A policy issued by the Federal Reserve which allows banks to keep only a small reserve of customer deposits on hand at any given time. They are free to loan out the rest of the funds to generate income through loans. Fractional reserve banking was eliminated on March 26, 2020.

Front-running: In crypto, a general term that means making moves in the market before everyone else. One example would be setting limit buy and sell orders at key levels of support and resistance.

FUD: Fear, Uncertainty, and Doubt. Used in reference to mainstream media's attempts to slander cryptocurrency.

Fundamental analysis: The process of determining the underlying strength of a project.

Fungible tokens: Tokens that are identical to one another, and can easily be exchanged and divided, just like paper money. Ex: BTC and ETH.

Game theory: A theory that claims that in a strategic setting, a participant's choice of action will depend on the actions of other participants. In the context of cryptocurrency, institutions are much more likely to invest knowing that their competitors are doing the same.

Genesis block: The first block that is mined for a Proof of Work cryptocurrency. The genesis block for bitcoin was mined on January 3rd, 2009 by Satoshi Nakamoto.

Golden cross: When a short term moving average crosses a longer moving average on the way up. An example would be the 50 day EMA crossing the 200 day EMA.

Halving: Bitcoin block rewards are cut in half every 4 years to taper the minting of bitcoin so that it steadily increases in value. As of May 11, 2020, miners receive 6.25 bitcoin per block. Also referred to as the halvening.

Hammer: A candle with a small body, little or no upper wick, and a long lower wick that is more than 2/3rd the length of the candle's total size. Hammers appear at or near the bottom of a correction or downtrend, and signal a potential reversal to the upside.

Hanging man: A hammer that appears after a green candle during an uptrend, which signals a potential reversal to the downside.

Hardware wallet: Any secure device that stores private keys offline. See also software wallet.

Hash: A unique fingerprint that identifies a block and its data. Bitcoin hashes data from each block into a 256-bit number using an algorithm known as

SHA256. Hashes are combined with those of previous blocks to form the blockchain.

Hiccup: In crypto, a pause where the price consolidates before making its next big leap.

HODL: An acronym for **H**old on **F**or **D**ear **L**ife, which began as a misspelling and turned into a full fledged investment strategy.

HODL fund: Your safety net, or a specific percentage of your investments that you won't sell for a predetermined length of time. This is often done to keep emotions out of the equation.

Hot wallet: See software wallet.

Initial Coin Offering (ICO): A method of raising funds for a project before launch that involves selling a percentage of pre-mined coins to early investors and venture capitalists, who often received a major discount for investing early. ICOs have mostly been replaced by IEOs and IDOs in recent years.

Initial Dex Offering (IDO): When a crypto start-up launches its token on a decentralized exchange to raise funding from retail investors.

Initial Exchange Offering (IEO): Fundraisers for crypto-startups that take place directly on centralized exchanges, which have the benefit of increased trust and transparency.

Initial margin: The collateral a trader offers in order to receive a loan from a participating exchange to trade using leverage. See also leverage trading.

Key pairs: A key pair for a cryptocurrency consists of both a private and public key, which are linked mathematically. The private key can reveal the public key, but it isn't possible to derive a private key from a public key.

Keychain: A more suitable term for a crypto wallet, since keychains hold your public/private keys, thereby giving you access to your crypto holdings. You can make copies of your keys and transfer them to as many keychains as you'd like. See also wallet.

Know Your Customer (KYC): The process of providing a government-issued identification, social security number, and home address for personal verification. KYC is mandated on regulated exchanges for the purposes of tax reporting. KYC requirements vary by exchange, and change frequently.

Lagging indicators: Most technical indicators, such as moving averages and RSI, are considered lagging, since they use price data that has already been recorded.

Leverage trading: Using debt in an attempt to amplify one's gains in a short time period. Using 10x leverage on a $100 investment means that you would gain $100 if the price of the asset went up 10%. If the price went down 10%, your position would be liquidated, and you would lose your entire investment. Also referred to as margin trading. See also initial margin.

Limit orders: Offers on an exchange's order books to buy and sell coins at specific prices. Creating a limit order adds liquidity to an exchange. For this reason, an investor who uses limit orders is referred to as a market maker.

Liquidation: When an exchange or trading platform closes a leveraged position after a trade loses more than the initial margin. A trader who chooses 100x leverage will be liquidated if the price drops 1% from the buy in price.

Liquidity: A measure of the ease of buying and selling an asset. Limit orders create liquidity, while market orders take it away.

Liquidity mining: See yield farming.

Liquidity pool: A collection of tokens locked in a smart contract, which are used to facilitate exchanges between different tokens.

Liquidity provider (LP): A crypto user who supplies liquidity to a pool, thereby earning a percentage of fees generated relative to his/her stake in the pool.

Loan to Value (LTV): A ratio of the percentage of funds borrowed against the value put up as collateral. Some lending protocols have a maximum LTV above which they liquidate your collateral.

Lockup period: The amount of time your crypto will be unavailable for trading after adding to a stake pool or lending platform.

Long position: Betting that an asset will rise in price.

Loss aversion: The human brain is wired to avoid the pain of loss instead of taking the necessary risk to win.

Margin trading: See leverage trading.

Market cap dominance: A percentage that shows the market cap of a cryptocurrency relative to the total market cap for the cryptosphere.

Market capitalization (market cap): The price of an asset multiplied by its circulating supply.

Market cycle: A period that typically lasts several months to years, and includes four stages: accumulation, markup, distribution, and decline. Conscious crypto investors time their entries during accumulation, make their exits during distribution, and sit on the sidelines during both markup and decline.

Market maker: An investor who adds liquidity to a market by placing limit orders, which helps to stabilize the price of an asset. Market makers pay lower fees than market takers, since they are essentially providing a service by adding liquidity to the order books. See also market taker.

Market order: An instant buy/sell order at current prices that removes liquidity from the market. Higher fees are paid in return for convenience and certainty that an order will be filled.

Market taker: An investor who uses a buy/sell market order to take away liquidity.

Markup: Stage two of a market cycle, which begins when a coin's price makes a clean breakout from its accumulation channel. An uptrend characterized by higher highs, higher lows, and a steady increase in buying volume. Moving averages will also begin to slope sharply upward during markup.

Marubozu: Japanese for "bald head." Candlesticks with little or no wick in either direction.

Maximalists (maxis): Cryptocurrency enthusiasts who are convinced of the superiority of original cryptos such as bitcoin and ethereum. Don't bother arguing with a Maxi!

Meme coins: Cryptocurrencies without any real world utility that usually rally around funny jokes or attention-grabbing sentiments. They often have a viral image plastered all over social media. Meme coins are the most manipulated coins in the market. Buyers beware!

Metcalfe's Law: The value of a project is proportional to its user base squared. More users on a network create more connections that increase the value of a cryptocurrency.

Micro-cap coins: Cryptocurrencies with less than a $100 million market cap. Considered to be the riskiest type of investment.

Mining: The process where computer nodes validate transactions on the blockchain to earn rewards in the form of freshly minted bitcoin (or another PoW crypto). Miners must first solve complex math puzzles to win the right to add the next block of transactions to the blockchain.

Minting: The process of creating a coin, usually either through Proof of Work or a pre-mine during a project launch.

Mobile wallet: See software wallet.

Moonboys: Crypto influencers who shamelessly shill their bags for their own personal gain.

Mooning: A term in cryptocurrency that refers to parabolic upward price movement. Ex: "SOL is mooning like crazy right now!"

Moving average: The average price of an asset over a given period of time, which is used to help recognize and clarify existing trends.

Network effect: In crypto, an ever-increasing number of projects that are eager and willing to use an existing blockchain network.

Newbs: Internet slang for retail investors who are new to crypto. Also known as normies.

NGMI: An acronym for **N**ot **G**onna **M**ake **I**t. A reference to crypto investors who make uninformed decisions. NGMI implies that certain mistakes will somehow lead you off a cliff and into the bottomless pit of fiat enslavement (they probably won't). If you are the victim of NGMI jokes, laugh and unfollow. See also YGMI.

Nocoiners: People who don't hold any cryptocurrency.

Nodes: Computers that join a cryptocurrency network. Nodes help to create consensus by verifying the accuracy of the blockchain.

Non-fungible tokens (NFTs): Tokens that represent unique or collectible items that aren't interchangeable, and can't be divided. Examples would be artwork or music, the deed to a house, or medical records. They make it possible to digitize assets, and are going to revolutionize numerous industries.

Non-custodial wallet: A wallet/keychain that is solely under an individual crypto user's custody. There is no outside authority that can access your funds, since they do not have access to your private keys. Non-custodial wallets are considered the most secure. See also custodial wallet.

Oracle coins: Cryptocurrencies that bring real world data to smart contract blockchains, thereby bridging the gap between the cryptosphere and the digital world.

Order books: A visual representation of ask/bid orders that can be viewed within an exchange's user trading panel. Green lines represent buy orders, while red lines represent sell orders. The current price falls in the middle.

Oscillators: Coins with unpredictable price movements that can be recognized by sharp, choppy waves on an expanded chart.

Pandemic of overexposure: Allocating too much of your resources into risky investments that you can't afford to lose.

Panic sell: Closing a trade immediately to prevent further loss.

Paper hands: See weak hands.

Parabolic: When the price of a cryptocurrency increases so rapidly that its chart forms a parabola.

Peer-to-peer (P2P): Person-to-person transactions that do not require 3rd party intermediaries such as banks.

Play-to-earn (P2E): Online gaming economies where players can earn rewards that are paid in cryptocurrencies. The tokens earned can either stay in-house or be cashed out. Axie Infinity is one such example, where players can breed Axies and sell them on the Axie marketplace.

Price action: The movement of an asset's price over time.

Price discovery: The idea that the prices of various cryptocurrencies will constantly be in flux as they are bought and sold. The collective perception of their value changes as new information enters the cryptosphere.

Privacy coins: cryptocurrencies that privatize transactions on the blockchain by hiding address balances and/or sender and recipient addresses.

Private key: An encrypted alphanumeric code, or a mathematical string of letters and numbers that grants access to holdings for a specific cryptocurrency. Private keys prove ownership of crypto, just like a key grants access to a lockbox. Not your keys, not your crypto! See also public key.

Proof of Stake (PoS): A cryptocurrency where nodes with the largest number of coins locked into the network validate transactions, eliminating the need for mining. This is said to be better for the environment, since significantly less electricity is used.

Proof of Work (PoW): The process of running computer software to solve complex mathematical puzzles to prove that transactions are legitimate. The effort in the form of time, resources, and electricity to solve the equations is the PoW that earns mining rewards.

Public address: A hashed version of your public key, which you use to receive crypto. A public address is often compared to an email address. See also public and private key, key pairs.

Public distributed ledger (PDL): A complete record of all the transactions that have ever occurred on a blockchain network.

Public key: Public keys are used to create public addresses where you can safely send and receive crypto without revealing your private key. Public keys are pseudonymous, since they are linked to you but don't reveal information about your personal identity. They are similar to transparent PO boxes where only you can receive mail. Anyone with your public address can track the movement of mail/crypto, but they can't access it without your private key.

Pumps and dumps: Rises and falls in price over time. When a crypto is rising fast, investors are "pumping" up the price. When it is plummeting, investors are "dumping" their supply into the market.

Range bound: See sideways trend.

Recency bias: Making assumptions based on information on a limited timeline.

Relative strength index (RSI): A momentum oscillator that compares the speed and change of a coin's price over the past 14 days. The RSI is a technical indicator that is popular among traders, and can be helpful for spotting divergences. See also bearish/bullish divergence.

Relief rally: A temporary price recovery during a larger downtrend.

Resistance: When the price of a coin rises to the same level and then drops back down again two or more times. A price ceiling.

Retracement: When the price of an asset moves back up toward previous highs after a correction.

Return on Investment (ROI): How much an investor gains from an initial investment. If you invested $100 in 1 BTC and then sold it at $10,000, your ROI would be $9,900.

Root seed: A very long number from which countless key pairs can be generated. Root seeds are translated into seed phrases.

Sacred pause: A moment of pure awareness, where you simply choose to be. Whatever you do to remember your eagle perspective, whether it be taking a series of deep breaths, a short hike in nature, or watching the sunset. Sacred pauses help us to be more alert and aware while making important trading decisions.

Satoshi Nakamoto: The alias of the creator(s) of bitcoin. Although many have speculated about the identity of Nakamoto, it remains unknown.

Sats: The smallest unit of bitcoin recorded on the blockchain: bitcoin to the eighth decimal. There are 100 million sats in 1 bitcoin. Sats are to bitcoin what cents are to the dollar, only much smaller.

Scalable: In terms of crypto, a blockchain with the potential to become faster and more efficient.

Scaling out: An expression for gradually selling a position. See also Dollar Cost Averaging.

Seed phrase: A series of 12 or 24 words that represent your root seed in plain english. Seed phrases are a gateway to your key pairs, and therefore your crypto holdings. If you lose your seed phrase, you lose access to your crypto. Sometimes referred to as a recovery phrase or a secret passphrase.

Selling into strength: Selling part or all of a position as the price of an asset is increasing.

Set it and forget it: Another way to say HODL. A long term investment strategy, buying and holding cryptocurrencies for several years.

Sheep: People who are either unwilling and/or unable to think for themselves, and follow both the opinions and decisions of the herd. They are easily manipulated by emotions such as fear and greed, since they haven't learned how to control them, and easily fall prey to market fluctuations. See also dumb money.

Shilling: The act of shamelessly promoting one's own investments to try and pump the price.

Short position: Betting that an asset will drop in price.

Sideways trend: A time period where the price stays in a parallel range with an upper and a lower boundary. Also known as consolidation, range bound.

Sizing: The process of determining how much money to invest in different cryptocurrencies. To a conscious crypto investor, appropriate sizing is one of the most important aspects of risk management.

Slashing penalties: A penalty to validator nodes that attempt to corrupt a PoS network. Validators must put up collateral in order to verify transactions on PoS networks. If they misbehave, they lose part or all of their collateral. This disincentivizes bad actors, since the risk outweighs the reward.

Slippage fee: A discrepancy in the price of two cryptocurrencies between the time a request is made and the swap is completed.

Smart contracts: Computer programs that perform a function automatically when a specific set of conditions are met. Smart contracts are immutable and decentralized–they can't be modified or shut down by a central authority.

Smart money: An attitude and way of being that is characterized by a clear, balanced mind that is able to keep the emotions in check, making wise investment decisions that will minimize risk and maximize gains over many years and decades. Becoming smart money is a life-long process that refines our character and benefits not only ourselves, but also the generations to come. We all have the potential to transform our attitude to become smart money.

Social tokens: Tokens issued by major brands that create a sense of inclusion by encouraging community participation. Companies will be able to leverage their massive followings by integrating social tokens into virtual hubs where fans can both interact and transact. Gaming tokens are a subset of social tokens, since they are centered around online gaming ecosystems.

Soft token: A time sensitive passcode generated by 2FA apps such as Authy that refreshes every 30 seconds. See also two factor authentication.

Software wallet: Any online application that stores your private keys, hence giving you access to your crypto. Also referred to as a hot wallet or mobile wallet.

Spread: The difference between the bid and ask price. Higher spreads mean market orders are more likely to fill at unexpected prices (either favorable or unfavorable).

Stablecoins: Cryptocurrencies with a value that is pegged to a fiat currency such as USD, another cryptocurrency, or a commodity such as gold. They are (supposed to be) backed by reserve assets, and are able to offer general price stability in an otherwise volatile market.

Stake pool: An address that is run by a validator, where delegators send their cryptocurrency to be staked on the network.

Staking: Locking crypto onto a network in return for rewards.

Stop loss: Sell orders automatically executed by a trading platform once the price drops to a preset level. Stop losses are used by traders to prevent further loss.

Store of Value Cryptocurrencies (SOV): A category of cryptos that either maintain or increase purchasing power over time. Bitcoin is considered a superior store of value because it is a deflationary asset with a fixed supply.

Super cycle: The theory that bitcoin and other cryptocurrencies will break out of their four year cycle, and that the cryptosphere will see much longer bull markets, followed by sharp, yet less pronounced bear markets.

Support: When the price of a coin drops to a certain area and then bounces up and away two or more times. A price floor.

Systematic correction: Marketwide stab wound corrections where most cryptocurrencies experience 30-90% dips. They often "bleed out" as they drop steadily over time, with aftershocks for some time after the initial trauma has passed. Depending on their severity, this type of correction can take weeks, months, or sometimes even years to fully heal.

Take profit: An order for the trading platform to automatically close part or all of a trade on your behalf at a predetermined price.

Technical analysis (TA): The study of charts to determine price trends.

The flippening: The belief that ethereum's market cap will one day surpass that of bitcoin.

The golden rule of investing: Your well-being is your top priority. Invest in yourself first.

The middle way: A calm, balanced approach to life that keeps us from leaning too far to any extreme.

The Tao: (pronounced Dao) The way, which involves harmony with the source of existence. Being an eternal student of the ebb and flow of Nature helps us to accept everything just as it is, without forcing or seeking to control.

Ticker: The symbol for a cryptocurrency. Ex: BTC for bitcoin.

Time preference: The value we place on receiving cash now vs. some time in the future. A lower time preference creates a more sustainable future for both people and planet.

TL;DR: An acronym for **T**oo **L**ong, **D**idn't **R**ead. The crypto version of an abstract, which contains a summary of the information presented in a whitepaper or other publication.

Token burn: The process of removing tokens from circulation by transferring them to an address that no one can access. Token burns have a deflationary effect that increases the price of the remaining supply.

Tokenomics: The details of a coin's creation, management, and distribution.

Tokens: Cryptocurrencies for projects that operate using protocols from existing blockchain networks. A crypto token is a type of coin.

Total Value Locked (TVL): The amount of money locked into the smart contracts of a blockchain network.

Trading bots: Computer programs that execute trades for you based on certain parameters that you set.

Trading channel: Parallel trend lines that reveal an asset's existing support and resistance levels. Channels can slope upwards, downwards, or range sideways.

Trading pair: Two assets that are traded together, such as BTC/USD or AVAX/USDT.

Transactions per second (tps): A measure of a blockchain's efficiency when it comes to processing transactions. TPS varies widely across different blockchains, and depends on the consensus mechanism, hash rate, and other factors.

Trend reversal: When an established trendline that connects a minimum of three wicks is broken. See also bullish and bearish reversal.

Trend trading: Timing exits and entries according to changes in the overall trend. The trend is your best friend!

Trendline: A line that connects the wicks of at least three candles, without cutting through any candle bodies.

Trustless: In crypto, blockchain networks use various decentralized consensus mechanisms that eliminate the need for trust. Miners and validators achieve consensus on all transactions, which prevents bad actors from tampering with the blockchain.

Two factor authentication (2FA): Additional pieces of information that identify you in order to keep your accounts secure. Examples of 2FA could be a PIN or an

SMS or email verification code. In the cryptosphere, 2FA generally involves using Authy or Google Authenticator to generate a unique soft token linked to your account that refreshes every 30 seconds.

Ultrasound money: Money that naturally lowers our time preference, inspiring us to delay gratification and plan for our future by saving and/or investing. Bitcoin, ethereum, and other cryptocurrencies are considered ultrasound money.

Uptrends: A price trend characterized by both higher highs and higher lows as the price steadily increases.

Validators: Node operators that verify and secure transactions on a Proof of Stake blockchain network to earn a portion of the fees.

Vaporware: A label for crypto projects that are still considered concepts in development, since they haven't yet implemented technology that can be utilized.

Vesting Schedule: The process of releasing coins into circulation over time. Ideally it is planned gradually, to prevent crashes from recipients dumping supply into the market.

Virtual private network (VPN): A network that allows you to set your IP address to anywhere in the world. VPNs use extra encryption to protect both your data and your identity.

Volume: The quantity of an asset that is bought and sold in a specified timeframe. Volume is a representation of the emotional intensity of investors, and can be used to either confirm or reject price movements.

Waiting On Awesome Happenings (WOAH): Investors who resist the urge to fall prey to FOMO, and instead do the opposite of the crowd. WOAH folks will wait patiently for prime investment opportunities, often buying when there's blood in the streets (everyone else is selling).

Wallet: In terms of crypto, a convenient, but poor choice for a crypto storage device or application. Crypto wallets don't actually store your funds–they hold the private keys which you can then use to access your crypto holdings. See also keychain.

Weak hands: Investors who are susceptible to selling their bags after dramatic price movements. Weak hands generally refers to newer, less experienced investors, but this is not always the case! Also known as paper hands.

Whales: Investors–either individuals or institutions–that own significant quantities of cryptocurrency. For bitcoin, a whale is considered to have more than 1,000 BTC.

White swan event: An unexpected positive event or announcement that leads to a tremendous surge in price of nearly every cryptocurrency over a very short time. Tesla's announcement that they had purchased $1.5 billion in BTC was a white swan event that caused the price of bitcoin to double in a few short weeks. See also black swan event.

Whitelisting: The practice of letting exchange(s) know which public addresses are safe to receive your crypto. You can save public addresses for specific cryptocurrencies, and then select from a dropdown menu in the "withdraw to" field. This saves the hassle of constantly having to copy/paste an address and verify its safety.

Whitepaper: A document that explains the purpose of a project and the problem(s) it proposes to solve.

Wolves: People who always enter the game knowing their strategy, and who don't allow emotion to lead them astray. When they act on their plans, they are swift and decisive. They are masters at blending in with their prey, keeping their presence unknown while maintaining proximity.

Wrecked (or rekt on CT): Losing large sums of money in crypto.

YGMI: An acronym for **Y**ou're **G**onna **M**ake **I**t. A cryptonaut who is deemed to be in the elite class of successful crypto investors by the twitter gods. At 1,000 YGMI approvals, you may receive a special plaque signed by one of Satoshi Nakamoto's representatives. See also NGMI.

Yield farming: Any form of using crypto to earn yield, including: lending, staking, and/or adding crypto to various liquidity pools. Also known as liquidity mining.

YOLO: Acronym for **Y**ou **O**nly **L**ive **O**nce. The best excuse on social media for aping into a crypto project and/or jumping out of a plane.

Index

C

D

E

F

G

H

I

K

L

M

N

O

P

Q

R

S

U

V

W

Y

Z

Made in United States
North Haven, CT
10 November 2026

22096541R00189